Microalgae Biotechnology - From Fundamental Research to Industrial Applications

Edited by Joana Campos

Published in London, United Kingdom

Microalgae Biotechnology - From Fundamental Research to Industrial Applications
http://dx.doi.org/10.5772/intechopen.1009716
Edited by Joana Campos

Assistant to the Editor: Raquel Fernandes

Contributors
Harish Bhandary, Homeira Zare Chavoshy, Joana Campos, Lamya Al-Musallam, Negar Nasri, Raquel Fernandes, Vanitha Vinod Kumar, Wid Alrashidi

First published in London, United Kingdom, 2026 by IntechOpen
IntechOpen is the global imprint of INTECHOPEN LIMITED, registered in England and Wales, registration number: 11086078, 167-169 Great Portland Street, London, W1W 5PF, United Kingdom

For EU product safety concerns: IN TECH d.o.o., Prolaz Marije Krucifikse Kozulić 3, 51000 Rijeka, Croatia, info@intechopen.com or visit our website at intechopen.com.

British Library Cataloguing-in-Publication Data
A catalogue record for this book is available from the British Library

Microalgae Biotechnology - From Fundamental Research to Industrial Applications
Edited by Joana Campos
p. cm.
Print ISBN 978-1-83634-554-1
Online ISBN 978-1-83634-553-4
eBook (PDF) ISBN 978-1-83634-555-8

Meet the editors

Joana Campos is a researcher with a multidisciplinary background that bridges microbiology, biotechnology, and project management. She completed her Ph.D. in Pharmaceutical Sciences at the University of Porto (UP) in 2018, specializing in the genomic epidemiology of multidrug-resistant bacteria. She further developed expertise in bioinformatics and human microbiome research at the Institute for Research and Innovation in Health (i3S-UP). She subsequently transitioned into applied biotechnology, working as a researcher and project manager in the food, cosmetics, and pharmaceutical industries, focusing on developing and valorizing microalgae-derived bioactive compounds, including participation in EU-funded projects. She is currently a researcher at the School of Medicine and Biomedical Sciences (ICBAS-UP).

Contents

Preface

Microalgae are among the most promising biological resources in modern biotechnology. Their rapid growth, metabolic plasticity, and biosynthesis of a wide range of high-value compounds have increased interest in using microalgae to develop innovative strategies to address global challenges in food production, human health, renewable energy, and environmental sustainability. *Microalgae Biotechnology – From Fundamental Research to Industrial Applications* brings together scientific contributions that explore both the current state of the art and future perspectives in this evolving field.

This book provides an integrated overview of the microalgae value chain, encompassing cultivation strategies, biomass production, and advanced biotechnological applications. The volume opens with the chapter *Brine to Bioresource: Transforming Brine Waste into Microalgal Wealth*, which extends the cultivation perspective by addressing sustainable production models that valorize saline waste streams within a circular-economy, resource-efficient framework. The second part of the book focuses on the biotechnological and biomedical relevance of microalgae-derived compounds, beginning with *Microalgae in Medicine: Emerging Therapeutic Applications*. This chapter gives an in-depth primer on the biological activities and therapeutic potential of microalgal metabolites, highlighting their relevance for human health. This thematic focus is further developed in *Phycocyanin as a Natural Antimicrobial for Biotechnological Development*, which examines a specific high-value compound in detail, addressing its antimicrobial properties, mechanisms of action, and translational potential across different biotechnological sectors. The book concludes with *Microalga-Based Electricity Production*, highlighting their integration into bioenergy systems and their relevance to the development of sustainable, forward-looking technological solutions.

Taken together, this book is intended to be a valuable reference for researchers, graduate students, and professionals working in biotechnology, environmental sciences, bioengineering, and related disciplines. By integrating fundamental knowledge with emerging technologies, *Microalgae Biotechnology – From Fundamental Research to Industrial Applications* seeks to advance sustainable biotechnological solutions by exploring the transformative potential of microalgae for next-generation bio-based industries.

Joana Campos
School of Medicine and Biomedical Sciences (ICBAS),
University of Porto,
Porto, Portugal

Raquel Fernandes
Centre for the Research and Technology of Agro-Environment
and Biological Sciences (CITAB),
University of Trás-os-Montes e Alto Douro,
Vila Real, Portugal

Section 1

Cultivation Strategies and Sustainable Production of Microalgae

Chapter 1

Brine to Bioresource: Transforming Brine Waste into Microalgal Wealth

Vanitha Vinod Kumar, Vinod Kumar, Lamya Al-Musallam and Harish Bhandary

Abstract

In recent years, seawater desalination technologies have advanced substantially, particularly in arid regions where the demand for freshwater is increasing. During the desalination process, brine is produced, which is characterized by elevated salinity, concentrated trace metals, residual chemical additives, and other dissolved contaminants. Unregulated disposal of large quantities of brine can severely disrupt marine ecosystems, causing shifts in physicochemical properties, loss of microbial diversity, and degradation of critical and sensitive habitats in the marine ecosystems. This chapter provides an overview of recent advances, integrating published findings and our experience on the sustainable utilization of reject brine for microalgae cultivation. Certain microalgal species belonging to genera *Dunaliella*, *Nannochloropsis*, *Chlorella*, and *Scenedesmus* not only tolerate high salinity but also produce valuable biomolecules, such as β-carotene, lipids, and proteins. This chapter provides insights into a selection of salt-tolerant microalgal species and cultivation strategies using brine, with an emphasis on the production of high-value biomolecules from brine-grown microalgae, while highlighting their simultaneous role in heavy metal sequestration and nutrient bioremediation. Comparative case studies and circular economy perspectives are explored to evaluate real-world applicability. The future directions emphasize further improvements in brine management through the development of suitable and designer microalgal strains using synthetic biology, genetic improvement, and integration of advanced precision monitoring systems for sustainable brine water management, especially in countries where desalination is the primary source of freshwater.

Keywords: brine water, desalination, protein, β-Carotene, microalgae, biofuel, *Dunaliella*

1. Introduction

1.1 Desalination and brine discharge

Desalination is widely used to produce potable water in arid and semiarid regions, particularly in the Middle East and North Africa (MENA). There are approximately 15,906 desalination plants that are operational worldwide, producing an estimated 96 million cubic meters (mm^3) of freshwater per day [1]. The desalination plants draw large amounts of brackish or seawater from diverse sources to meet the freshwater demand. A major environmental concern associated with desalination is the generation and disposal of brine, also referred to as "reject," "reject brine," or "desalination concentrate" [1–3]. Brine is a concentrated saline effluent of desalination and has a salinity level approximately twice that of seawater, with elevated levels of chemicals including antiscalants, coagulants, and residual disinfectants. On an average, for every liter of freshwater production, about 1.5–1.7 liters of brine is produced [1]. The desalination plants are typically located along the coastal regions and hence the produced brine is commonly discharged back into the marine environment. Such discharges pose significant ecological risks due to high salinity and the potential accumulation of chemical contaminants and toxic chemicals. A lack of adequate circulation and dilution of brine further enhances the deleterious effects.

The chemical composition of brine generated varies depending on the composition of feed water and the desalination process. The majority of desalination processes employ either thermal or membrane-based technologies. Thermal technologies primarily encompass multi-effect distillation (MED) and multi-stage flash (MSF) distillation [1, 2, 4]. These systems rely on phase change and heat input to separate freshwater from saline feed water. In contrast, membrane-based technologies utilize semipermeable membranes to separate salts from saline water. Among these, Reverse Osmosis (RO) is the most widely adopted technology, accounting for approximately 74% of global desalination capacity. Thermal technologies contribute around 24%, while the remaining 2% is attributed to electrodialysis (ED) and electrodialysis reversal (EDR) [2, 4]. The Middle East accounts for nearly 50% of the globally installed desalination capacity and corresponding brine discharge (**Table 1**), with the world's largest desalination plant situated in Saudi Arabia [2].

1.2 Characteristics of brine water

Brine water discharged from desalination plants contains elevated levels of salts, chemicals, and many contaminants. Brine generated from seawater desalination typically exhibits salinity levels ranging between 50 and 80 ppt, which is nearly double

Region	Number of desalination plants	Desalination capacity (million m^3.day^{-1})	Brine production (million m^3.day^{-1})
Middle East & North Africa	4826	45.32	99.4
East Asia & Pacific	3505	17.52	14.9
Western Europe	2337	8.75	8.4
North America	2341	11.34	5.6
Latin America & Caribbean	1373	5.46	5.6
Eastern Europe & Central Asia	566	2.26	2.5
Southern Asia	655	2.94	3.7
Africa	303	1.78	1.5
Total	**15,906**	**95.37**	**141.5**

Table 1.
Global desalination capacity and brine production [1].

the average salinity of natural seawater (~35 ppt). The physicochemical composition of brine is highly variable and depends on multiple factors, including the desalination technology employed (e.g., thermal or membrane-based), the quality of intake water (brackish/ground/seawater), and the local environment and contamination levels [5]. In addition to common ions: sodium (Na^+), chloride (Cl^-), magnesium (Mg^{2+}), calcium (Ca^{2+}), and sulfate (SO_4^{2-}), brine also contains residual chemicals, such as phosphate-based antiscalants, sulfate-based coagulants, trace amounts of metals such as copper, zinc, lead, pH adjusters, and disinfectants (i.e., chlorine) [1, 6]. These factors strongly influence the quality of the discharged brine (**Table 2**), leading to regional variability in composition and concentration across different countries [5]. In addition, pronounced seasonal heterogeneity is observed in the brine composition. For instance, brines from the Persian/Arabian Gulf desalination plants often exhibit higher salt concentration, compared to the Mediterranean countries, mainly based on the source water utilized for desalination [5].

1.3 Environmental impacts of brine disposal

Desalination plants employ various brine disposal methods, including direct discharge to the ocean, sewer discharge, evaporation ponds, surface water discharge, deep well injection, and land application [2, 4]. Most desalination plants are located along coastal zones where an improper disposal of brine discharge can trigger a cascade of ecosystem-level consequences through multiple environmental pathways (**Figure 1**). Discharged brine infiltrates terrestrial systems, contributing to soil salinization, diminished crop yields, structural degradation of soil, and long-term impact on agriculture. Percolation into subsurface layers introduces contaminants and heavy metals, affecting the quality of groundwater. The use of brine water for irrigation leads to an increase in soil salinity, which leads to decreased plant productivity. The deleterious effect of brine on agricultural crops, salt deposition, and a potential use

Parameter	Average range	Unit
Salinity	50–80	ppt
pH	6.5–8.5	—
Temperature	28–42	°C
Na^+ (Sodium)	0.3–25	$g.L^{-1}$
K+	32–950	$mg.L^{-1}$
Cl^- (Chloride)	35–50	$g.L^{-1}$
Mg^{2+} (Magnesium)	17–2974	$mg.L^{-1}$
Ca^{2+}	14–1180	$mg.L^{-1}$
SO_4^{2-} (Sulfate)	2.5–4.0	$g.L^{-1}$
Total dissolved solids (TDS)	1410–81,492	$mg.L^{-1}$
Cu^{2+} (Copper)	0.05–0.3	$mg.L^{-1}$
Zn^{2+} (Zinc)	0.01–0.1	$mg.L^{-1}$
Residual chlorine	0.1–0.5	$mg.L^{-1}$

Table 2.
Physicochemical properties of brine water. These data represent the data ranges from multiple countries [1].

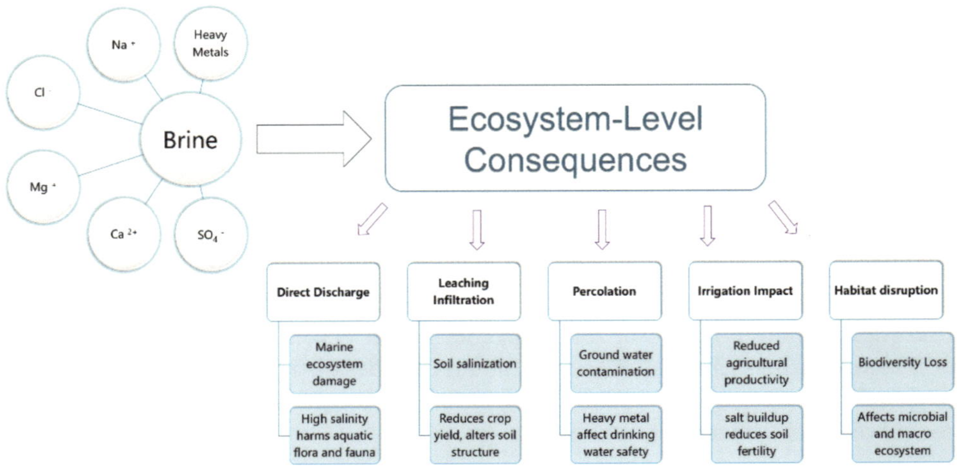

Figure 1.
Constituents of brine and potential effects on ecosystem upon disposal.

for sustainable food production in greenhouse and hydroponics has been reviewed with a case study from Qatar [7].

Discharge of brine enriched with salts (Na^+, Cl^-, Mg^{2+}, Ca^{2+}, and SO_4^{2-}) and heavy metals can damage marine ecosystems, salinize soils, contaminate groundwater, and disrupt habitats, ultimately leading to biodiversity loss and altered ecosystem functioning.

In the marine environment, the direct release leads to substantial ecosystem damage, disrupting the aquatic flora and fauna. The high salinity and thermal load of brine effluents produce localized salinity and temperature regions/plumes in the receiving marine environment, which disrupt natural physicochemical gradients and osmotic balance, affecting plankton communities, coral assemblages, and benthic fauna. These plumes can extend from as close as 15 meters from the discharge point to distances exceeding 5 kilometers, depending on the discharge volume, hydrodynamic conditions, and seafloor topography [5, 8]. In addition, brine frequently contains trace metals, such as copper, zinc, and lead, along with chemical additives, chlorine, and antiscalants. These can accumulate in sediments and pose chronic toxicity risks to marine organisms and bioaccumulate in food chains (**Figure 2**).

Desalination brine containing elevated salinity, heavy metals, residual oxidants, antiscalants, and biocides enters the marine ecosystem, where limited dispersal promotes accumulation. Contaminants could affect species diversity, affect interactions at multiple levels within the community, enter the food web at lower trophic levels, and undergo bioaccumulation and biomagnification, leading to imbalances in energy flow, shifts in community composition, and overall ecosystem degradation.

The combined effects of chemical and salinity stress can reduce microbial diversity near discharge zones, thereby altering biogeochemical cycles and lessening ecosystem resilience [2, 9]. Documented biological impacts of brine discharge include developmental abnormalities in fish and shrimp larvae, tissue degeneration and increased mortality, reduced fertilization in sea urchins, coral bleaching, and reduced polyp formations. Although the seagrass species, *Posidonia australis*, demonstrated tolerance to mild salinity fluctuations, higher salinity was detrimental to the growth of seedlings [1, 2, 10]. The chemicals accumulated on the seabed form a layer that disturbs the sessile benthic invertebrates, such as bryozoans (*Microporella* sp., *Smittina* sp.,

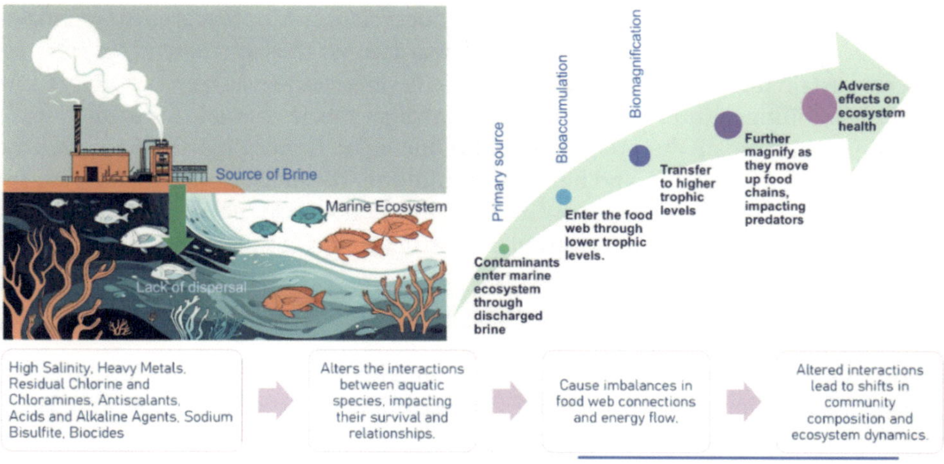

Figure 2.
Brine discharge impact on the marine environment.

Calloporina sp.), sponges, and kelp (*Ecklonia radiata*), polychaetes, and echinoderms, leading to a marked population decline [11]. The elevated copper and nickel exhibited localized mangrove mortality in Ras Hunjurah lagoon in the UAE [12], illustrating species-specific vulnerabilities of marine organisms to brine-related stressors. Habitat alteration caused by saline and sediment-laden effluent can lead to fragmentation and loss of native species. These impacts are especially pronounced in enclosed or shallow water bodies, such as the Persian/Arabian Gulf. However, a case study from Spain over a period of 17 years of long-term monitoring found that pre-dilution of brine prior to discharge can significantly minimize the ecological impact on the brine discharge area [13]. The impacts of brine discharge are due to the cumulative effects of discharge volumes, poor dilution, low flushing and hydrodynamically confined areas, and seasonal stagnation leading to progressive accumulation of salts and other chemicals.

1.4 Sustainable brine management

Given the ecological and environmental impacts associated with brine disposal, there is a strong need for sustainable brine management strategies. Mitigation measures such as discharging brine into the ocean or injecting it into deep wells merely shift the problem spatially, without addressing the inherent environmental risks. To avoid brine discharge altogether, resource valorization (the process of converting waste to value) techniques must be deployed. Numerous halophilic bacterial species have been identified in hypersaline marine environment and these can be cultivated using brine as a growth medium. A comprehensive review of fungi and bacteria capable of growth in brine environments has been compiled by Rotter et al. [11].

Cultivation of halotolerant microalgae offers a promising path to mitigate the deleterious effects of brine [5, 11, 14, 15]. Halotolerant microalgae can withstand high salinities, utilize excess nutrients present in brine, and yield commercially valuable biomaterials, including carotenoids, proteins, and lipids. Biological treatment systems can also adsorb critical trace metals and neutralize residual chemical additives, reducing harm to the environment. These three outcomes of nutrient recovery, nutrient recycling (as usable biomass), and bioremediation of harmful waste outputs address traditional aspects of pollution, as well as stimulate future resource recovery and recycling. This evolution of contamination, waste, and byproduct relationships represents a circular bioresource paradigm and turns brine from hazardous waste to an increasingly strategic input for sustainable biotechnology. Further in this chapter, we discuss the opportunities and possibilities to utilize salt-tolerant microalgal species suitable for brine water cultivation and evaluate their potential for producing high-value bioproducts, such as carotenoids, lipids, proteins, and biofuels. In addition, the mechanisms underlying the tolerance to salt stress, environmental and techno-economic feasibility, and integrating the microalgae systems into desalination infrastructure for circular resource recovery.

1.5 Microalgae as a sustainable solution

Microalgae represent a viable and environmentally sustainable option for the productive utilization of brine discharge from desalination processes. These photosynthetic autotrophic microorganisms are capable of thriving in diverse habitats, including hypersaline environments, making them ideal for cultivation in brine effluents. Species, such as *Dunaliella salina*, *Scenedesmus quadricauda*, *Nannochloropsis* sp., and *Chlorella vulgaris*, have demonstrated robust growth under high salinities while producing commercially valuable biomolecules, including β-carotene, astaxanthin, lipids, and proteins.

Cultivating microalgae using desalination brine not only reduces the negative effect of saline effluent discharge on the environment but also utilizes available carbon dioxide (CO_2), thereby potentially contributing to the reduction of greenhouse gas emissions. This approach simultaneously produces biomass with applications across biofuels, nutraceuticals, cosmetics, and animal feed industries. Additionally, certain microalgae have the ability to bioaccumulate nutrients and trace metals, contributing to the partial bioremediation of the brine, prior to its final disposal.

Thus, integrating microalgae cultivation into brine management supports the circular economy principles and aligns with relevant sustainable development goals (SDGs 6, 12, and 13) by converting waste effluent into a valuable renewable bioresource. Depending on the design and desalination infrastructure, brine-based algal cultivation can be implemented in closed or semi-closed loop systems, potentially reducing the use of freshwater and minimizing environmental discharge impacts. Furthermore, recent advances in biotechnologies, including but not limited to genetic engineering and Artificial Intelligence (AI) monitoring technologies, offer significant opportunities to enhance the scalability, productivity, and economic potential of algal systems that utilize brine as a feedstock. This integrated approach will provide a sustainable pathway for managing saline waste, particularly in areas with high desalination and brine discharge.

2. Microalgal species cultivated in brine water

Microalgae possess remarkable tolerance to high-salinity levels, allowing certain species to thrive in brine water. These species can grow in extreme ionic conditions, while simultaneously producing high-value bioproducts such as β-carotene, lipids, and proteins under salinity-induced stress. Hence, the cultivation of microalgae utilizing brine water as a growth medium provides an economical and environmental friendly desalination option.

Microalgae belonging to the genera *Dunaliella, Nannochloropsis, Scenedesmus, Spirulina*, and *Chlorella* are widely used in research studies. *D. salina* is the most experimented microalga due to its high-salinity tolerance and its remarkable capacity for β-carotene accumulation at salinities ranging from 0 to 110 ppt [16, 17].

Algae sp.	Biomass yield	Bioproducts	Reference
N. gaditana	0.96 $g.L^{-1}$ in 75% brine	Lipids: 4.7–12.6% Proteins: 27%	[15, 18, 19]
C. vulgaris	Up to 0.6 $g.L^{-1}$ in 25% brine	Lipids: 12.5% Proteins: 45.3–51.0%	[18, 24]
S. platensis	1.55 $g.L^{-1}$ in 50% brine	Lipids: 4.5% Proteins: 50%	[14, 19]
D. tertiolecta	0.2–0.3 $g.L^{-1}$ (Concentrated and 50% brine)	—	[19, 25]
N. oculata	0.2 $g. L^{-1}$ (Concentrated and 50% brine)		[25]
S. quadricauda	Not specified	High protein biomass	[5, 19, 26]
Dunaliella (sp. BCA421-ITC5.003)	26.44 $g.m^{-2}.day^{-1}$	Nutrient removal from brine	[21]
D. salina, KU11	0.33 $g.L^{-1}$ (brine)	Beta-carotene – 19 $mg.L^{-1}$	[27]
D. salina	2.5 $g.L^{-1}$	Beta-carotene – 2.5 $mg.L^{-1}$	[23]
Phaeodactylum tricornutum, Tetraselmis suecica, Pavlova lutheri	Not specified	—	[28]
C. vulgaris	1.49–2.5 $g.L^{-1}$	Lipids: 15%	[29]

Table 3.
Microalgal species cultivated in brine water and productivity.

Nannochloropsis gaditana and *C. vulgaris* produced substantial biomass at salinities of 30–60 ppt, producing lipids for conversion to biodiesel [15, 18]. In a few case studies, the reject was mixed with conventional media, as some of the species did not grow with the concentrated brine. For *C. vulgaris*, 25% dilution was found optimal for the growth [15, 19], whereas *Spirulina* exhibited growth in 50% brine [19, 20]. Zhu et al. (2018) pretreated the brine to precipitate 97% of Ca^{2+} and 60% of Mg^{2+}, followed by CO_2 enrichment to achieve a bicarbonate-rich medium for the growth of *D. salina*. In a few other studies, the brine water was enriched with nitrogen and phosphorus [21, 22] to provide a nutrient balance for the effective growth of *Dunaliella* species.

However, in our studies, concentrated brine with a salinity of 65 ppt, for the cultivation of *D. salina*, exhibited double the growth rate compared to the seawater-based media of 35 ppt [23]. Hence, microalgal species can be cultivated using brine, either directly or after appropriate pretreatment or following dilution with the required amount of freshwater, depending on the tolerance of the selected species. Optimizing biomass yield and biochemical composition requires a careful selection of species based on salinity tolerance, nutrient availability, and specific production objectives—such as biomass yield, pigment, lipid, or protein composition (**Table 3**).

3. Bioproducts from microalgae

Microalgae are recognized as a valuable source of high-value products, producing lipids, proteins, carotenoids, carbohydrates, and polyunsaturated fatty acids. Lipids have significant importance for biodiesel production [30]. Carotenoids such as beta-carotene are potent antioxidants, with commercial relevance in pharmaceuticals, health food, and cosmetics [31]. Proteins provide essential amino acids and are suitable for aquafeed and functional foods [32]. Polyunsaturated fatty acids support cardiovascular and neurological health [33]. Carbohydrates are important for biopolymer production [34]. Under stress conditions, *D. salina, S. platensis*, and *Nannochloropsis* accumulate pigments, such as β-carotene, astaxanthin, and phycocyanin, along with lipids and long-chain fatty acids valuable for pharmaceutical and nutraceutical applications.

4. Nutrient removal from brine

In addition to biomass production and valuable products, microalgae cultivation can significantly contribute to sustainable and effective means of bioremediation of brine by sequestering salts, heavy metals and removing excess nutrients. There are numerous studies on the utilization of freshwater microalgae for sewage wastewater treatment to clean polluted water and to remove heavy metals [35, 36]. Similarly, species such as *D. salina, Picochlorum, Nannochloropsis*, and *C. vulgaris* have high tolerance to salinity and are capable of accumulating salts and cations, such as Mg^{2+}, Ca^{2+}, and Na^{+} [5, 37]. During the active growth stages, the concentrations of nitrate, phosphate, sulfate, and other nutrients were reduced in the media [5].

Bioremediation efficiencies of microalgae during cultivation are primarily dependent on brine pH, light intensity, brine chemistry, and the tolerance threshold of the selected species [5, 37]. *Scenedesmus* was found to remove around 83% of nitrogen and phosphorus [6], whereas *Picochlorus* utilized only 50% [5]. *D. salina* was known to reduce 79–99% of nitrate and 12–71% of phosphate across different seasons [21, 38].

Microalgal species	Type of brine	Nutrient removal	Efficiency	Reference
Spirulina	Brine	Nitrate, phosphate	96% nitrate, 83% phosphate removal	[39]
Scenedesmus	Brine	Nitrate, phosphate	100% nitrate removal 83% phosphate removal	[40]
P. tricornutum	Brine	Nitrate	Nitrate removal	[28]
Dunaliella (sp. BCA421-ITC5.003)	Brine	Nitrate, phosphate	99% nitrate removal 71% phosphate removal	[21]
D. tertiolecta, N. oculata	Concentrated and 50% brine	Nitrate, phosphate	Nitrate, phosphate removal	[25]
D. salina	Brine	Nitrate, phosphate	Nitrate, phosphate removal	[41]
Nannochloropsis	Brine	Nitrogen	99% total nitrogen removal 67% phosphate removal	[42]
D. salina	Brine	Nitrate	95% nitrate removal 87% fluoride removal	[23]

Table 4.
Use of reject brine for the nutrient remediation by microalgae.

The bioremediation efficiencies of microalgae vary from study to study and may depend on various factors. Under optimized conditions, the reported studies have shown a 70–90% reduction of nitrate and phosphate levels from the brine media (**Table 4**). Most of the studies used diluted or pretreated brine to balance the nutrient composition and salinity. However, *Dunaliella* species are well suited due to their adaptation to high salinity and their ability to produce beta-carotene under salt stress conditions. Hence, microalgae bioremediation is a potentially viable and sustainable approach of integrated brine management in aquaculture and desalination systems.

5. Genetic tolerance mechanisms and metabolomic adaptations under hypersaline conditions

Some species of microalgae have developed genetic mechanisms and metabolomic adaptations to survive and reproduce in hypersaline brine. Cellular synthesis of osmoprotectants, such as glycerol, proline, trehalose, sorbitol, and mannitol, helps to maintain osmotic balance and contributes to their survival mechanisms [17]. In *Dunaliella* species, the intracellular levels of glycerol increase in relation to the external salt concentration. This mechanism helps in balancing the osmotic pressure within the cells and prevents cell lysis [43, 44]. Another notable balance of osmotic pressure and osmotic lysis is maintained by the regulation of ion transporters. The key ion transporters like Na^+/H^+ antiporters and H^+-ATPases actively pump sodium ions out of the cell or sequester them into vacuoles [16, 45, 46]. The balance of cytoplasmic ion concentrations is critical for enzymatic functions in microalgae.

Stress-inducible genes are reported to have roles in anaerobic and high-salinity responses. Key mechanisms of cell survival can be attributed to the role of heat shock proteins (HSPs) as protectors of protein integrity under thermal and osmotic stress. In addition, superoxide dismutase (SOD) and catalase (CAT) groups of enzymes effectively quench reactive oxygen species (ROS), which could otherwise damage cellular components and cell membrane when exposed to high levels of salinity [47]. The regulation and expression of stress response genes in microalgae are tightly regulated by transcriptional networks, which are activated upon salinity sensing (**Figure 3**).

Increased salinity stress triggers activation of cellular signal transduction pathways in microalgae. In a model organism, *Dunaliella*, two major responses are initiated: (i) glycerol biosynthesis, which supports osmoprotection, and (ii) ion transporter activation, including Na^+/H^+ antiporters and H^+-ATPases, which mediate sodium sequestration or efflux. These processes result in the activation of stress-responsive genes such as heat shock proteins (HSPs), superoxide dismutase (SOD), and catalase (CAT), and other key regulators enhancing the salt tolerance of the microalgae. With these mechanisms, halotolerant species exhibit improved cell survival, sustained multiplication, and produce secondary metabolites under hypersaline stress conditions.

With the advent of transcriptomics and CRISPR (Clustered Regularly Interspaced Short Palindromic Repeats) tools, researchers can now investigate stress regulatory genes such as heat shock factor A1 (HsfA1), Salt Overly Sensitive (SOS) pathways, and gene cluster transporters to gain unique and valuable engineering opportunities for high-salinity tolerance [47, 48]. Alterations in metabolomics most often include upregulation of antioxidant molecules, such as β-carotene [16, 49], lipids [29, 50], and other secondary metabolites. Furthermore, salt stress results in alterations to metabolic pathways and increases the production of fatty acids, polyphenols, and

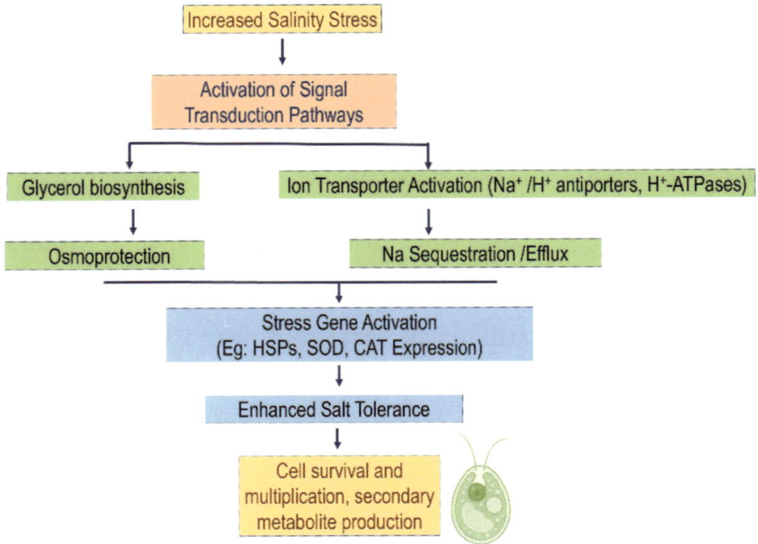

Figure 3.
Cellular response to salinity stress in halotolerant microalgae.

protective pigments, such as β-carotene and astaxanthin [47]. The levels of fatty acids such as palmitic acid (C16:0) and oleic acid (C18:1) increased 20–40% when subjected to salt stress in *Dunaliella* [51]. Under salt stress, the protein content decreased, while the lipid content increased, compared to conventional media. Collectively, these adaptations enable microalgae to maintain membrane stability, preserve photosynthetic efficiency, and mitigate ion-induced cytotoxicity under osmotic stressors.

6. Integrated systems for brine management

This section illustrates a schematic representation of a conceptual framework that integrates desalination plants, microalgae cultivation systems, and downstream utilization of algal biomass for biomass production and bioremediation. The desalination process generates large amounts of brine, and instead of discharging them directly into the marine environments, they can be used as a potential source of nutrients for microalgae farms. However, redirecting the brine to bioreactors or open ponds requires a careful consideration of monitoring and readjustment of physicochemical parameters to support the growth of microalgae. During the growth phase, the salt-tolerant microalgae can assimilate the residual nutrients and also absorb dissolved CO_2 in the brine, leading to the production of valuable resources while capturing a significant amount of CO_2.

The model can be further improved by incorporating renewable energy for the operation of algae reactors/ponds, sensor-based monitoring, and biomass harvesting systems. The system could be advanced further by minimizing the disposal of effluents by reuse of the medium for multiple culture production cycles. Matos et al. [15, 19] conducted three successive cultivation cycles of *N. gaditana* with nutrient supplementation.

As depicted in the conceptual diagram (**Figure 4**), the brine serves as both a nutrient and a water source for algal cultivation. The generated biomass can be

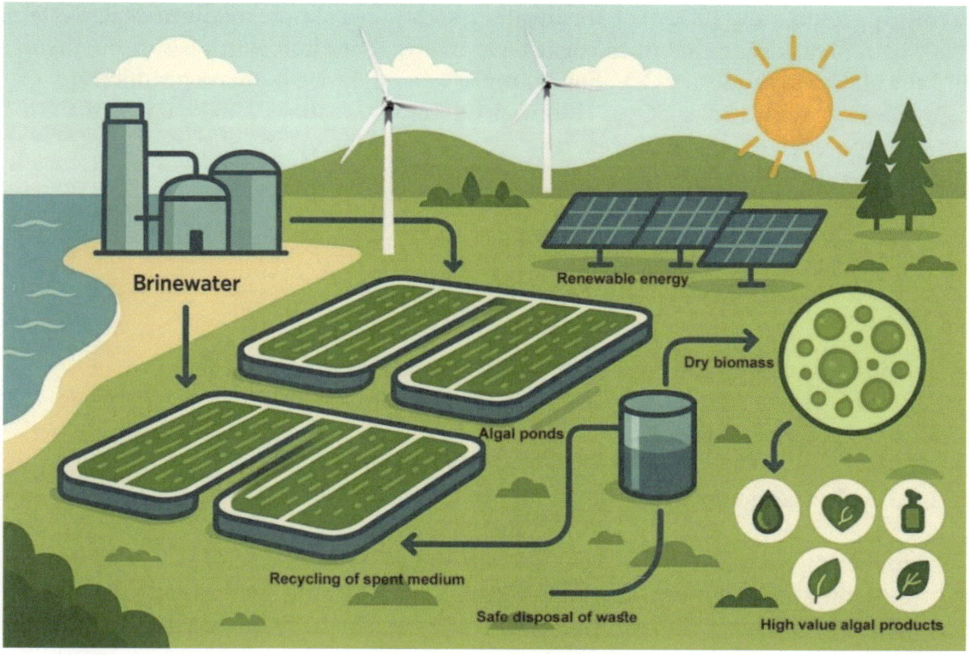

Figure 4.
Integrated brine-algae-desalination system.

further processed into protein (food/feed), lipid (biofuel source), or other high-value compounds such as carotenoids. Simultaneously, the microalgae contribute to the treatment of brine water, and the treated water can be recycled again with media nutrient supplementation or discharged back with minimal environmental impact.

7. Environmental and economic impacts

Microalgae grown in brine systems offer substantial environmental and economic benefits. Environmentally, the brine-based algae farming system can minimize marine pollution through the reduction of brine being disposed of directly into the coastal environment while utilizing the inherent waste streams. Economically, the brine cultivation process converts saline waste into high-value bioproducts. In summary, brine used promotes circular economy systems that reduce waste, diversify income, and also provide a use for land that may not be suitable for crop production in the desert arid ecosystem. Thus, the proposed platform offers a noncompetitive alternative to food crops by utilizing nonarable resources, aligning with the increasing industrial demand for low-cost, environmentally compatible solutions in bioremediation and green biotechnology commercialization.

7.1 Sustainability metrics

Three primary sustainability issues are addressed by microalgae cultivation using brine: freshwater overuse, carbon emissions, and marine pollution. Reusing brine in microalgae cultivation reduces its water footprint by substituting brine water for that of freshwater. In practical terms, 1 m^3 of reused brine replaces 1 m^3 of freshwater, saving approximately 1000 L of freshwater per day, in a simple medium-scale system. It is worth mentioning that microalgae have an efficient photosynthesis mechanism and the ability to sequester CO_2. For example, at a reference biomass productivity of 1.2 $g.L^{-1}.day^{-1}$ with a 1.8 g $CO_2.g^{-1}$ biomass fixation rate, about 2.16 $kg.day^{-1}$ of CO_2 sequestration occurs in 1000 L reactors.

Equation: CO_2 fixed per day.

$$CO_2\text{Fixed} = \text{Biomass Yield}\left(\text{g.L}^{-1}.day^{-1}\right) \times CO_2\text{Fixed } per \text{ gram Biomass} \times \text{Volume}\left(\text{L}\right) = 1.2 \times 1.8 \times 1000 = 2160\,\text{g}.day^{-1} = 2.16\,kg.day^{-1} \quad (1)$$

In addition, during the growth, microalgae can significantly reduce nutrient pollution and some species can absorb heavy metal ions within a closed-loop eco-restorative output framework.

7.2 Life cycle assessment

Conducting a detailed Life Cycle Assessment (LCA) for microalgae that are grown in brine compared to conventional freshwater systems shows significantly lower environmental impacts. Substituting freshwater with brine in microalgae cultivation can reduce water consumption by approximately 80%, while also lowering energy demand by eliminating the need for daily pumping and freshwater purification. Based on cradle-to-gate LCA, produced biomass grown in brine emits 0.8 kg CO_2-eq. kg^{-1} biomass relative to 1.6 kg CO_2-eq. kg^{-1} for conventional systems. Brine systems account for reduced eutrophication potential from nutrient uptake. Overall, these reductions make microalgal brine cultivation climate-smart and circular because they contribute to the United Nations' Sustainable Development Goals to clean water, sustainable city sustainability, and responsible consumption.

8. Future prospects and research directions

8.1 Genetic engineering for enhanced tolerance and yields

The potential for using genetic engineering to enhance microalgal productivity in hypersaline environments is exciting. Gene editing tools such as CRISPR-Cas9 (CRISPR-associated protein 9) could allow for the integration of salt-tolerance genes, including Na^+/H^+ antiporters. In addition, genes involved in the biosynthesis of osmo-protectants (glycerol, proline, glycine, betaine) and lipid biosynthesis can be targeted to improve stress adaptation and metabolite accumulation. Transgenic strains, such as *D. salina*, overexpressing antioxidant gene pathways, could offer improvements of 30–50% in biomass yield and enhance carotenoid accumulation over unmodified strains [52]. These modifications could help to reduce the lag phase of growth rates, as well as ionic stress tolerance, helping to develop more resilient and reliable brine-based cultivation.

8.2 Synthetic biology and metabolic pathway engineering

Synthetic biology enables metabolic pathways and activities in microalgae to be reprogrammed for highly valuable products from brine substrates. Synthetic operons and optimized gene clusters redirect metabolic fluxes toward enhanced production of β-carotene, astaxanthin, or biofuels. Dynamic metabolic control (inducible promoters, biosensors) ensures efficient metabolic flux re-routing under saline stress or nutrient shifts [53]. Recent studies [54–59] have extensively explored the use of genetic engineering to develop improved microalgal strains with enhanced stress tolerance, demonstrating significant potential for customized industrial-scale applications.

8.3 AI- and sensor-based monitoring in brine cultivation

Optimizing microalgae production efficiency requires effective monitoring of growth rates and concentration in photobioreactor systems. Recent studies have proposed an automated and cost-effective solution utilizing machine learning (ML) with image-based analysis to overcome the limitations of traditional, resource-intensive methods [60]. ML-driven imaging offers a scalable, noninvasive, and real-time alternative for industrial-scale microalgae cultivation monitoring, which could be useful for algae grown in brine at an industrial scale. Artificial Intelligence (AI) combined with real-time sensors creates predictive control and adaptive feedback control for brine microalgae systems. The sensors can be utilized to evaluate pH, temperature, electrical conductivity, dissolved oxygen and CO_2, and irradiance, while AI algorithms can forecast potential growth trends and nutrient depletion [61]. This could potentially improve the efficiency of the overall process by 25–40%, while also reducing reliance on human capital and ensuring steady biological yield [62].

8.4 Integration with desalination plants and smart cities

Future city models do not just consider desalination plants as simply a source of sustainable freshwater, but instead, as sources of organic bioproducts. Through the integration of microalgae reactors and brine outlets, bioremediation can occur in situ while harvesting other bioproducts that utilize this brine discharge. The waste energy and CO_2 produced during the desalination of seawater can also be used in conjunction with any existing waste algal biomass to produce additional products from bioproducts [63]. Establishing solar-powered sensor suites reduces maintenance overhead and enables remote deployment of monitoring systems at commercial raceway ponds and thin-layer reactors [61]. Urban agri-tech systems with algae farms can be efficiently managed and monitored through AI- and Internet of Things (IoT)-based platforms, enabling integration and reuse of resources [60, 61, 64].

8.5 Gaps in knowledge and multidisciplinary opportunities

While improvements have been made, there remain several foundational gaps in knowledge surrounding brine-algal production systems. For example, currently there is no deep species-level/genomics-level understanding of halophilic microalgae and how they behave under fluctuating halophilic conditions. There has been little work done using multi-omics approaches to understand the epigenetic and evolutionary adaptations.

Salinity adaptation in microalgae also depends on the dynamic interactions with associated microbial consortia. The microalgal microbiome, comprising bacteria, archaea, and fungi residing in the phycosphere, plays a significant role in modulating algal stress responses [65]. The microbial partners contribute to adaptations by producing osmoprotectants and healing the algae in modulating ion transport, and facilitating nutrient uptake under saline conditions. Studies have shown that cocultivation of microalgae with specific halotolerant bacterial strains improves biomass productivity and photosynthetic efficiency under high salinity [65]. This field of research is still evolving and suggests that developing a designer consortium is a viable strategy to enhance the resilience of microalgal cultures for production in brine effluents from desalination processes.

Although the brine can serve as a growth medium for halotolerant species, research on its application in large-scale cultivation remains limited. Scaling up brine-based algal cultivation requires addressing practical barriers beyond strain tolerance and media chemistry. Availability of suitable species is limited, and intellectual property rights may restrict access to optimized strains. Only a few species can tolerate undiluted brine, whereas diluted brine supports the growth of a broader range of taxa. Large-scale growth also faces additional challenges, including the contaminant microorganisms that persist in high salinity and reduce the productivity. In some cases, antiscalant residues present in brine may precipitate during algal growth, leading to cell mortality or impaired growth. Furthermore, the concentrations of trace metals, toxicants, and other dissolved constituents in brine vary substantially across regions depending on feed water quality. Operational challenges include variation in brine composition that hinders process optimization. Therefore, standardization of brine-based culture media is critical for developing reliable large-scale cultivation methodologies. Managing salinity in the final disposal of spent brine after repeated recycling remains a critical environmental concern, especially with risks of xenobiotic accumulation. High capital investment for infrastructure, coupled with ongoing maintenance costs, adds further constraints to commercial deployment. Uncertainties in local and global demands for algal biomass products, and the need for skilled technical staff to maintain culture systems, remain a significant challenge that can affect the profitability in brine-based biomass production.

Furthermore, the long-term ecological implications of using brine for biomass production need to be assessed along with economic feasibility. Furthermore, the life cycle and techno-economic analysis studies remain spatiotemporally unique and lack any replication/adherence; therefore, it is hard to get a global scale. A multidisciplinary collaborative approach would offer numerous opportunities in this field, with biotechnologists able to engineer robust strains of halophilic microalgae. Concurrently, engineers can design adaptable modular photobioreactors, while data scientists can contribute by building predictive AI for optimizing algal yields in brine. Furthermore, urban planners have a role in incorporating these systems into sustainable smart city planning. Additionally, combining microalgae systems with microbial consortia, desalination rejects could help to boost sustainability and lengthen the value chain. Filling the gaps with innovation, in partnerships, will result in the significant potential of converting waste brine into a bioresource.

9. Conclusion

Converting hypersaline brine from a waste liability to a bioresource through microalgae cultivation is a promising intersection of environmental biotechnologies and circular economies. In this book chapter, we have outlined how halotolerant and halophilic microalgae can survive and thrive in brine, and contribute to removing nutrients, heavy metals, and other contaminants. Furthermore, the production of various bioproducts (e.g., protein, β-carotene, and lipids) adds an economic dimension to environmental stewardship. Life Cycle Assessment demonstrates that brine-based cultivation systems can reduce water and carbon footprints, offering sustainable alternatives to freshwater-based cultivation from both economic and sustainability practices. Advances in genetic engineering and synthetic biology, and the introduction of AI-based monitoring and control systems will further optimize brine-based production and increase resilience and yield at scale. Additionally, prospective integration scenarios with desalination plants and smart infrastructure could enable meaningful brine-based production. Nevertheless, challenges remain, particularly concerning economic feasibility, standardized systems, regulatory frameworks, and closing knowledge gaps in marine microbial ecology and long-term impacts of brine-based productions. Multidisciplinary approaches will be vital in addressing these knowledge hurdles. Overall, "brine-to-bioresource" is a cutting-edge paradigm toward sustainable resource management and offers new avenues for biotechnological evolution and innovation.

Acknowledgements

The authors gratefully acknowledge the support by Grant PR18-12SL-13 from the Kuwait Foundation for the Advancement of Sciences (KFAS) and additional support from the Kuwait Institute for Scientific Research (FM114C). The authors also acknowledge the support from Dr. Faiza Al Yamani, Dr. Mohsen Al Hussaini, and Dr. Turki Al Said for providing necessary support during the course of the study.

Conflict of interest

The authors declare no conflict of interest.

Author details

Vanitha Vinod Kumar[1*], Vinod Kumar[1], Lamya Al-Musallam[1] and Harish Bhandary[2]

1 Environment and Life Sciences Research Center, Kuwait Institute for Scientific Research, Kuwait

2 Water Research Center, Kuwait Institute for Scientific Research, Kuwait

*Address all correspondence to: vvkumar@kisr.edu.kw

References

[1] Jones E, Qadir M, van-Vliet MTH, Smakhtin V, Kang SM. The state of desalination and brine production: A global outlook. Science of the Total Environment. 2019;**657**:1343-1356

[2] Panagopoulos A, Haralambous K-J. Environmental impacts of desalination and brine treatment-challenges and mitigation measures. Marine Pollution Bulletin. 2020;**161**:111773

[3] Ghernaout D. Desalination engineering: Environmental impacts of the brine disposal and their control. Open Access Library Journal. 2020;7(9):1-17

[4] Panagopoulos A, Haralambous K-J, Loizidou M. Desalination brine disposal methods and treatment technologies-A review. Science of the Total Environment. 2019;**693**:133545

[5] Gilani IE, Hosseini H, Al-Ghouti MA, Saadaoui I, Sayadi S. Microalgal-based desalination brine remediation: Achievements, challenges, and future research trends. Environmental Technology & Innovation. 2024;**34**:103592

[6] Mezher T, Fath H, Abbas Z, Khaled A. Techno-economic assessment and environmental impacts of desalination technologies. Desalination. 2011;**266**(1-3):263-273

[7] Salsabila N, Al-Ansari T, Bicer Y. A review of the potential of seawater brine for enhancing food security in hot arid climates: A case study of Qatar. Journal of Environmental Management. 2025;**375**:124216

[8] Riera R, Tuya F, Ramos E, Rodríguez M, Monterroso Ó. Variability of macrofaunal assemblages on the surroundings of a brine disposal. Desalination. 2012;**291**:94-100

[9] Sirota R, Winters G, Levy O, Marques J, Paytan A, Silverman J, et al. Impacts of desalination brine discharge on benthic ecosystems. Environmental Science & Technology. 2024;**58**(13):5631-5645

[10] Cambridge M, Zavala-Perez A, Cawthray G, Mondon J, Kendrick G. Effects of high salinity from desalination brine on growth, photosynthesis, water relations and osmolyte concentrations of seagrass *Posidonia australis*. Marine Pollution Bulletin. 2017;**115**(1-2):252-260

[11] Rotter A, Rinkevich B, Deniz I, Reddy MM, Girão M, Carvalho MF, et al. Innovative solutions for valorization of desalination brine. Water Research X. 2025;**28**:100372

[12] Omerspahic M, Al-Jabri H, Siddiqui SA, Saadaoui I. Characteristics of desalination brine and its impacts on marine chemistry and health, with emphasis on the Persian/Arabian gulf: A review. Frontiers in Marine Science. 2022;**9**:845113

[13] Sola I, Fernández-Torquemada Y, Forcada A, Valle C, del Pilar-Ruso Y, González-Correa JM, et al. Sustainable desalination: Long-term monitoring of brine discharge in the marine environment. Marine Pollution Bulletin. 2020;**161**:111813

[14] Matos ÂP, da Silva T, Sant'Anna ES. The feasibility of using inland desalination concentrate (DC) as an alternative substrate for *Spirulina platensis* mass cultivation. Waste and Biomass Valorization. 2021;**12**(6):3193-3203

[15] Matos ÂP, Feller R, Moecke EHS, Sant'Anna ES. Biomass, lipid productivities and fatty acids composition of marine *Nannochloropsis gaditana* cultured in desalination concentrate. Bioresource Technology. 2015;**197**:48-55

[16] Oren A. A hundred years of Dunaliella research: 1905-2005. Saline systems. 2005;**1**:1-14

[17] Chen H, Jiang JG. Osmotic responses of *Dunaliella* to the changes of salinity. Journal of Cellular Physiology. 2009;**219**(2):251-258

[18] Matos ÂP, Saldanha-Corrêa FMP, Hurtado GR, Vadiveloo A, Moheimani NR. Influence of desalination concentrate medium on microalgal metabolism, biomass production and biochemical composition. Journal of Chemical Technology & Biotechnology. 2024;**99**(2):321-329

[19] Matos ÂP, Moecke EH, Sant'Anna ES. The use of desalination concentrate as a potential substrate for microalgae cultivation in Brazil. Algal Research. 2017;**24**:505-508

[20] Volkmann H, Imianovsky U, Furlong EB, Oliveira JLB, Sant'Anna ES. Influence of desalinator wastewater for the cultivation of *Arthrospira platensis*. Fatty acids profile. Grasas y aceites. 2007;**58**(4):396-401

[21] Ortega Méndez JA, Mendoza H, Santiago DE, Aridane Rodríguez F, Gil Lodos M, Carmona L. Reuse of SWRO brine for the production of carotenoids from *Dunaliella salina* and removal of macronutrients. Desalination and Water Treatment. 2012;**49**(1-3):115-122

[22] Yildirim O, Tunay D, Ozkaya B. Reuse of sea water reverse osmosis brine to produce *Dunaliella salina* based β-carotene as a valuable bioproduct: A circular bioeconomy perspective. Journal of Environmental Management. 2022;**302**:114024

[23] Vanitha VK, Al Musallam L, Kumar V, Habeebullah SFK, Bhandary H, Surendraraj A, et al. Mass Production and Evaluation of Extraction Methods for Efficient Recovery of Beta-Carotene from *Dunaliella sp*. Project Final Report, Environment and Life Sciences Research Center, Kuwait. Kuwait Institute for Scientific Research; 2021

[24] Matos Â, Ferreira W, Morioka L, Moecke E, França K, Sant'Anna E. Cultivation of *Chlorella vulgaris* in medium supplemented with desalination concentrate grown in a pilot-scale open raceway. Brazilian Journal of Chemical Engineering. 2018;**35**(4):1183-1192

[25] Shirazi SA, Rastegary J, Aghajani M, Ghassemi A. Simultaneous biomass production and water desalination concentrate treatment by using microalgae. Desalination and Water Treatment. 2018;**135**:101-107

[26] Al Bazedi G, Ismail MM, Mugwanya M, Sewilam H. Desalination concentrate microalgae cultivation: Biomass production and applications. Sustainable Water Resources Management. 2023;**9**(4):108

[27] Wu Z, Dejtisakdi W, Kermanee P, Ma C, Arirob W, Sathasivam R, et al. Outdoor cultivation of *Dunaliella salina* KU 11 using brine and saline lake water with raceway ponds in northeastern Thailand. Biotechnology and Applied Biochemistry. 2017;**64**(6):938-943

[28] Ridley CJ, Parker BM, Norman L, Schlarb-Ridley B, Dennis R, Jamieson AE, et al. Growth of microalgae using nitrate-rich brine wash from the water industry. Algal Research. 2018;**33**:91-98

[29] Almutairi AW, El-Sayed AE-KB, Reda MM. Evaluation of high salinity adaptation for lipid bio-accumulation in the green microalga *Chlorella vulgaris*. Saudi Journal of Biological Sciences. 2021;**28**(7):3981-3988

[30] Chisti Y. Biodiesel from microalgae. Biotechnology Advances. 2007;**25**(3):294-306

[31] Lamers PP, Janssen M, De Vos RC, Bino RJ, Wijffels RH. Exploring and exploiting carotenoid accumulation in *Dunaliella salina* for cell-factory applications. Trends in Biotechnology. 2008;**26**(11):631-638

[32] Becker EW. Micro-algae as a source of protein. Biotechnology Advances. 2007;**25**(2):207-210

[33] Ward OP, Singh A. Omega-3/6 fatty acids: Alternative sources of production. Process Biochemistry. 2005;**40**(12):3627-3652

[34] Ho S-H, Chen C-Y, Chang J-S. Effect of light intensity and nitrogen starvation on CO2 fixation and lipid/carbohydrate production of an indigenous microalga *Scenedesmus obliquus* CNW-N. Bioresource Technology. 2012;**113**:244-252

[35] Yan C, Qu Z, Wang J, Cao L, Han Q. Microalgal bioremediation of heavy metal pollution in water: Recent advances, challenges, and prospects. Chemosphere. 2022;**286**:131870

[36] Leong YK, Chang J-S. Bioremediation of heavy metals using microalgae: Recent advances and mechanisms. Bioresource Technology. 2020;**303**:122886

[37] Hasnain M, Zainab R, Ali F, Abideen Z. An overview of biosorption, remediation, and desalination potential of algae to remove wastewater toxic pollutants for environmental safety. Environmental Monitoring and Assessment. 2025;**197**(8):853

[38] Zhu C, Zhai X, Jia J, Wang J, Han D, Li Y, et al. Seawater desalination concentrate for cultivation of *Dunaliella salina* with floating photobioreactor to produce β-carotene. Algal Research. 2018;**35**:319-324

[39] Mata SN, de Souza ST, Cardoso LG, Andrade BB, Duarte JH, Costa JAV, et al. Spirulina sp. LEB 18 cultivation in a raceway-type bioreactor using wastewater from desalination process: Production of carbohydrate-rich biomass. Bioresource Technology. 2020;**311**:123495

[40] Mehdipour N, Hosseini SA, Hedayati A, Zolfaghari M, Kordi H. Determination of suitable compatible microalgae for bioremediation of effluent from desalination facilities. Biomass Conversion and Biorefinery. 2024;**14**(24):31233-31244

[41] de Souza CG, de Souza Schneider RdC, Rizzetti TM, Lobo EA, Sui Y. Using wastewater as a cultivation alternative for microalga *Dunaliella salina*: Potentials and challenges. Science of the Total Environment. 2024;**911**:168812

[42] ElBarmelgy A, Ismail MM, Sewilam H. Biomass productivity of *Nannochloropsis* sp. grown in desalination brine culture medium. Desalination and Water Treatment. 2021;**216**:306-314

[43] Ramos AA, Polle J, Tran D, Cushman JC, Jin E-S, Varela JC. The unicellular green alga *Dunaliella salina* Teod. as a model for abiotic stress tolerance: Genetic advances and future perspectives. Algae. 2011;**26**(1):3-20

[44] Goyal A. Osmoregulation in *Dunaliella*, part I: Effects of osmotic

stress on photosynthesis, dark respiration and glycerol metabolism in *Dunaliella tertiolecta* and its salt-sensitive mutant (HL 25/8). Plant Physiology and Biochemistry. 2007;**45**(9):696-704

[45] Ben-Amotz A, Avron M. On the factors which determine massive β-carotene accumulation in the halotolerant alga *Dunaliella bardawil*. Plant Physiology. 1983;**72**(3):593-597

[46] Shetty P, Gitau MM, Maróti G. Salinity stress responses and adaptation mechanisms in eukaryotic green microalgae. Cells. 2019;**8**(12):1657

[47] Panahi B, Hejazi MA. Weighted gene co-expression network analysis of the salt-responsive transcriptomes reveals novel hub genes in green halophytic microalgae *Dunaliella salina*. Scientific Reports. 2021;**11**(1):1607

[48] Neofotis P, Huang A, Chang W, Joseph F, Polle JE. Microalgae strain isolation, screening, and identification for biofuels and high-value products. In: Microalgal Production for Biomass and High-Value Products. Boca Raton, FL, USA: CRC Press; 2016. pp. 63-89

[49] Ben-Amotz A, Avron M. On the mechanism of osmoregulation in Dunaliella. In: Caplan SR, Ginzburg M, editors. Energetics and Structure of Halophilic Microorganisms, Developments in Halophilic Microorganisms. Vol. 1. Elsevier/North-Holland Biomedical Press; 1978. pp. 529-541

[50] Abomohra AE-F, El-Naggar AH, Alaswad SO, Elsayed M, Li M, Li W. Enhancement of biodiesel yield from a halophilic green microalga isolated under extreme hypersaline conditions through stepwise salinity adaptation strategy. Bioresource Technology. 2020;**310**:123462

[51] Ben-Amotz A. Bioactive compounds: Glycerol production, carotenoid production, fatty acids production. In: The Alga *Dunaliella*. CRC Press; 2019. pp. 189-208

[52] Radakovits R, Jinkerson RE, Darzins A, Posewitz MC. Genetic engineering of algae for enhanced biofuel production. Eukaryotic Cell. 2010;**9**(4):486-501

[53] Wehrs M, Tanjore D, Eng T, Lievense J, Pray TR, Mukhopadhyay A. Engineering robust production microbes for large-scale cultivation. Trends in Microbiology. 2019;**27**(6):524-537

[54] Amendola S, Kneip JS, Meyer F, Perozeni F, Cazzaniga S, Lauersen KJ, et al. Metabolic engineering for efficient Ketocarotenoid accumulation in the green microalga *Chlamydomonas reinhardtii*. ACS Synthetic Biology. 2023;**12**(3):820-831. Epub 2023/02/24. DOI: 10.1021/acssynbio.2c00616

[55] Shelake RM, Khalid MA, Kim JY. Microalgal bioengineering for futuristic applications in synthetic and space biology. Biotechnology Advances. 2025;**84**:108665. Epub 2025/07/31. DOI: 10.1016/j.biotechadv.2025.108665

[56] De Silvio MA, Sanchez-Retuerta C, Aguila Ruiz-Sola M, Baidukova O, Monte E. A quick-to-implement and optimized CRISPR-Cas9 protocol to obtain insertional and small indel mutants in *Chlamydomonas reinhardtii*. MethodsX. 2025;**15**:103416. Epub 2025/07/10. DOI: 10.1016/j.mex.2025.103416

[57] Tran QG, Le TT, Choi DY, Cho DH, Yun JH, Choi HI, et al. Progress and challenges in CRISPR/Cas applications in microalgae. Journal of Microbiology. 2025;**63**(3):e2501028. Epub 2025/04/08. DOI: 10.71150/jm.2501028

[58] Li Y, Deng L, Walker EJL, Karas BJ, Mock T. Genetic engineering in diatoms: Advances and prospects. The Plant Journal. 2025;**121**(6):e70102. Epub 2025/03/16. DOI: 10.1111/tpj.70102

[59] Calatrava V, Gonzalez-Ballester D, Dubini A. Microalgae for bioremediation: Advances, challenges, and public perception on genetic engineering. BMC Plant Biology. 2024;**24**(1):1261. Epub 2024/12/28. DOI: 10.1186/s12870-024-05995-5

[60] Uguz S, Sahin YS, Kumar P, Yang X, Anderson G. Real-time algal monitoring using novel machine learning approaches. Big Data and Cognitive Computing. 2025;**9**(6):153. DOI: 10.3390/bdcc9060153

[61] Soon CF, Jamia'an M, Mohamed Sunar N, SNH A, Tay KG, Heng C, et al. Smart sensing and anomaly detection for microalgae culture based on LoRaWAN sensors and LSTM autoencoder. Aquaculture International. 2025;**33**(6):438. DOI: 10.1007/s10499-025-02104-7

[62] Le-Thi-Thu N, Nguyen-Van-Duong T, Doan-Van T, editors. Combining artificial intelligence with the internet of things in microalgae cultivation. In: International Conference on Intelligence of Things. Cham: Springer International Publishing; 2022

[63] Al-Hotmani O, Al-Obaidi M, John YM, Patel R, Mujtaba I. Optimisation of hybrid MED-TVC and double reverse osmosis processes for producing different grades of water in a smart city. Desalination. 2022;**534**:115776

[64] Sumanasekara H, Jayasingha H, Amarasooriya G, Dayarathne N, Mainali B, Senevirathna L, et al. Hybrid machine learning models for predicting the impact of light wavelengths on algal growth in freshwater ecosystems. Phycology. 2025;**5**(2):23

[65] Wang T, Li D, Tian X, Huang G, He M, Wang C, et al. Mitigating salinity stress through interactions between microalgae and different forms (free-living & alginate gel-encapsulated) of bacteria isolated from estuarine environments. Science of The Total Environment. 2024;**926**:171909

Section 2

Microalgae-Derived Bioactive Compounds: Extraction, Functionalization and Bioactivity

Chapter 2

Microalgae in Medicine: Emerging Therapeutic Applications

Negar Nasri and Homeira Zare Chavoshy

Abstract

Microalgae, as prolific producers of bioactive metabolites, have emerged as promising candidates in the development of novel therapeutic agents. These unicellular photosynthetic organisms synthesize a diverse array of primary and secondary metabolites—including carotenoids, phycobiliproteins, polyunsaturated fatty acids (PUFAs), sulfated polysaccharides, and phenolic compounds—that exhibit potent pharmacological activities. Accumulating evidence from preclinical models and early-stage clinical trials supports their anti-inflammatory, antioxidant, immunomodulatory, anticancer, neuroprotective, and cardioprotective properties. Compounds such as astaxanthin, fucoxanthin, C-phycocyanin, and eicosapentaenoic acid (EPA)/docosahexaenoic acid (DHA) have demonstrated the ability to modulate critical signaling pathways, including NF-κB, Nrf2, MAPK, and JAK/STAT, thereby attenuating pro-inflammatory cytokine production, reducing oxidative stress, and inhibiting apoptosis in various disease models, including cancer, cardiovascular dysfunction, and neurodegenerative disorders. In parallel, sulfated polysaccharides and sulfolipids extracted from microalgae have shown immunoregulatory functions through macrophage activation, regulation of cytokine profiles, and interaction with innate immune receptors such as TLRs and Dectin-1. Advances in biotechnological tools, including high-throughput screening, omics technologies, and green extraction methodologies, have significantly facilitated the identification and production of these bioactive compounds. However, challenges related to standardization, scalability, pharmacokinetics, and regulatory validation remain major obstacles in the clinical translation of microalgae-based therapeutics.

Keywords: microalgae, bioactive compounds, primary and secondary metabolites, novel therapeutic agents, pharmacokinetics

1. Introduction

Microalgae are a diverse group of unicellular photosynthetic organisms found in a wide range of aquatic environments (**Figure 1**). Despite their microscopic size, they possess remarkable biosynthetic capabilities, producing an array of bioactive compounds, including proteins, polyunsaturated fatty acids, pigments, vitamins, and polysaccharides [1]. These bioactive molecules not only support the organisms' survival under extreme environmental conditions but also offer significant benefits for human health. In recent decades, the scientific community has increasingly

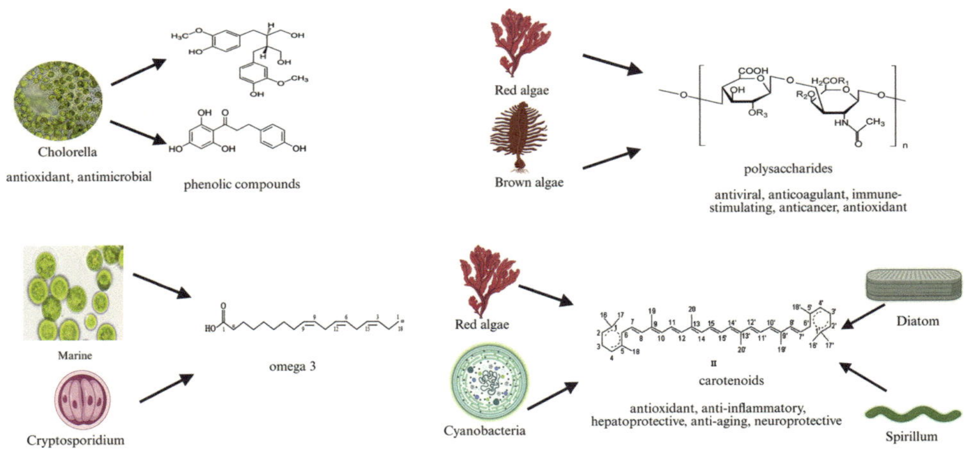

Figure 1.
Schematic of microalgae category and bioactive compounds.

focused on microalgae as a sustainable source of high-value products for nutraceutical, pharmaceutical, and cosmeceutical applications, owing to their rapid growth rates, minimal land use requirements, and carbon fixation potential [2].

The medicinal potential of microalgae is not a novel concept. Historical records and ethnopharmacological evidence indicate that certain microalgal species, such as *Spirulina* and *Chlorella*, have been used in traditional diets and healing practices across various cultures [3]. Notably, *Spirulina platensis* has garnered global attention for its rich nutrient profile and demonstrated immunomodulatory, antioxidant, and anti-inflammatory properties [4]. Over time, advancements in biotechnology have facilitated the identification and extraction of specific therapeutic compounds from microalgae, transitioning their use from traditional remedies to evidence-based medical applications. This evolution underscores a growing recognition of microalgae not merely as dietary supplements but also as potential sources of novel therapeutics [5].

This chapter aims to explore the emerging therapeutic applications of microalgae in modern medicine by examining current scientific findings and ongoing research. It will delve into the pharmacological properties of microalgal metabolites and their roles in the prevention and treatment of various human diseases, including metabolic disorders, cancer, and infectious diseases. Furthermore, the chapter will highlight challenges associated with the development of microalgae-based therapies and propose future directions for their integration into clinical practice. By synthesizing insights from biotechnology, pharmacology, and molecular biology, the discussion seeks to underscore the transformative potential of microalgae as an innovative and sustainable resource in therapeutic development.

2. Diversity and biochemistry of microalgae

2.1 Taxonomic diversity and habitat

Microalgae represent a remarkably heterogeneous yet ecologically indispensable group of photosynthetic organisms found within both prokaryotic and eukaryotic domains (**Figure 2**). These organisms, though unified by their ability to conduct oxygenic photosynthesis, are evolutionarily disparate and exhibit vast taxonomic

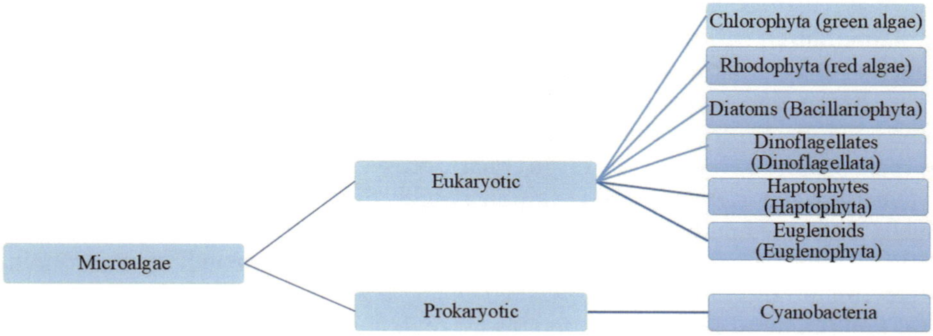

Figure 2.
Microalgae taxonomy.

diversity [6]. Microalgae are not a monophyletic group; rather, they have arisen through multiple independent endosymbiotic events, including primary, secondary, and even tertiary endosymbioses, which have given rise to a wide spectrum of chloroplast structures, pigment profiles, and metabolic configurations [7].

Among prokaryotes, *cyanobacteria* are the most prominent photosynthetic representatives. Formerly known as "blue-green algae," *cyanobacteria* possess thylakoid membranes with chlorophyll a and unique accessory pigments such as phycocyanin and phycoerythrin. Though they lack membrane-bound organelles, they carry out oxygenic photosynthesis and are considered evolutionary ancestors of modern chloroplasts. Notably, certain filamentous *cyanobacteria* possess specialized nitrogen-fixing cells called heterocysts, enabling them to thrive in nitrogen-depleted environments. Their ability to produce structurally diverse secondary metabolites, including neurotoxins and UV-absorbing compounds, has attracted increasing attention in both ecological and biotechnological research [8].

In the eukaryotic realm, microalgal diversity is even more pronounced. Eukaryotic microalgae represent a remarkably diverse group of aquatic photoautotrophs, characterized by a wide range of structural, pigmentary, and metabolic adaptations [9]. This diversity arises largely from the complex evolutionary history of plastid acquisition via primary and secondary endosymbiosis. During primary endosymbiosis, a primitive eukaryote engulfed a cyanobacterium, giving rise to plastids with two membranes, as seen in groups such as *Chlorophyta* (*green algae*) and *Rhodophyta* (*red algae*). Secondary endosymbiosis involved a eukaryotic host assimilating a photosynthetic eukaryotic alga, resulting in plastids surrounded by three or four membranes—an evolutionary event that underpins groups such as diatoms, *dinoflagellates*, *haptophytes,* and *cryptophytes* [10].

Among the most ecologically and biotechnologically relevant lineages, *Chlorophyta* (*green algae*) stand out due to their close evolutionary relationship with terrestrial plants [11]. These organisms possess chlorophylls *a* and *b*, store starch within their chloroplasts, and often feature cellulose-based cell walls. Their chloroplasts may assume diverse morphologies, including cup-shaped and star-shaped forms. While predominantly found in freshwater ecosystems, members of *Chlorophyta* can also inhabit soil surfaces, tree bark, and even snow, exhibiting a high degree of ecological flexibility. Their rapid growth rates, ease of cultivation, and capacity to produce bioactive compounds make them prime candidates for bioindustrial applications [12].

Diatoms (*Bacillariophyta*), another major group of eukaryotic microalgae, are unicellular organisms encased within intricately patterned silica-based cell walls known as frustules. These walls not only provide protection but also serve as a basis for taxonomic classification due to their unique nanostructures. *Diatoms* possess chlorophylls *a* and *c* and the accessory pigment fucoxanthin, giving them a characteristic golden-brown hue. Found in marine, freshwater, and even moist terrestrial environments, diatoms contribute significantly to global primary production and are among the most prolific oxygen producers in aquatic ecosystems [13, 14].

Dinoflagellates (*Dinoflagellata*) exhibit a highly distinctive morphology with two flagella positioned perpendicularly within grooves of their cellulose-based theca. This group displays immense metabolic and ecological versatility, including autotrophic, heterotrophic, and mixotrophic species. *Dinoflagellates* are key components of marine planktonic communities and are notable for their ability to form harmful algal blooms (HABs), some of which produce potent biotoxins. Additionally, certain species are bioluminescent, emitting light in response to mechanical stimuli—a phenomenon commonly observed in tropical oceans [15, 16].

Haptophytes (*Haptophyta*) are predominantly marine microalgae that are recognized for their unique cellular appendage, known as the haptonema, and their ability to biomineralize calcium carbonate into elaborate plate-like structures called coccoliths. These structures play an essential role in the biological carbon pump, facilitating carbon sequestration in ocean sediments. *Haptophytes* contain chlorophylls *a* and *c*, along with fucoxanthin, and are capable of producing long-chain polyunsaturated fatty acids such as EPA and DHA, which are of growing interest in nutraceutical and aquafeed industries [17].

Euglenoids (*Euglenophyta*) are a morphologically and metabolically flexible group that bridges characteristics of both plants and *protozoa*. Lacking a rigid cell wall, they possess a proteinaceous pellicle that allows for cellular plasticity and movement. Euglenoids are mixotrophic, able to photosynthesize in the presence of light and absorb organic compounds in its absence. These adaptive traits enable them to thrive in fluctuating environments, particularly in stagnant freshwater systems rich in organic matter. *Euglena gracilis*, a well-studied species, has garnered attention for its potential in biofuel production and recombinant protein expression. Overall, eukaryotic microalgae inhabit an astonishing variety of ecological niches—from freshwater lakes, rivers, and wetlands to marine pelagic zones and extreme environments, such as acidic springs, hypersaline ponds, polar ice, and desert soils. Their adaptive versatility and evolutionary innovations not only underpin their ecological significance but also render them invaluable resources in modern biotechnology, medicine, and sustainable development [18].

3. Therapeutic compounds from microalgae

3.1 Antioxidants (e.g., astaxanthin, β-carotene)

Among the most prominent and extensively studied biological activities of microalgal metabolites is their antioxidant potential, which plays a crucial role in mitigating oxidative stress—a pathological state driven by the excessive accumulation of reactive oxygen species (ROS) and reactive nitrogen species (RNS). Oxidative stress has been implicated in the etiology of numerous chronic conditions, including cardiovascular diseases, neurodegenerative disorders, cancer, diabetes, and persistent inflammation. Microalgae synthesize a broad spectrum of bioactive compounds with notable antioxidant properties, including carotenoids (e.g., astaxanthin, lutein, β-carotene), phycobiliproteins (e.g., phycocyanin), sulfated polysaccharides, polyphenols, and omega-3 fatty acids, each operating through distinct yet often complementary mechanisms. The primary mode of action for many of these compounds involves direct free radical scavenging [19]. Their molecular structures—particularly the presence of conjugated double bonds and functional groups capable of donating hydrogen or electrons—enable them to neutralize ROS and RNS efficiently. For instance, astaxanthin's extended polyene chain and keto groups confer exceptional radical-quenching capacity, often surpassing traditional antioxidants like vitamin C or E. This radical-scavenging activity protects critical biomolecules such as lipids, proteins, and nucleic acids from oxidative damage, thereby preserving cellular function and structural integrity [20].

In addition to these direct effects, several microalgal compounds modulate intracellular redox signaling pathways, particularly the Nrf2–Keap1–ARE axis, which governs the transcription of genes involved in antioxidant defense. Under basal conditions, the transcription factor Nrf2 is sequestered in the cytoplasm by its inhibitor Keap1. Upon exposure to oxidative stress or stimulation by specific microalgal metabolites, Nrf2 dissociates from Keap1, translocates into the nucleus, and binds to antioxidant response elements (ARE) in the promoter regions of target genes. This cascade induces the expression of endogenous antioxidant enzymes such as superoxide dismutase (SOD), catalase, and glutathione peroxidase (GPx), which collectively enhance the cell's ability to neutralize ROS and restore redox balance [21]. Some microalgal compounds, such as phycocyanin, have also been shown to inhibit the enzymatic sources of ROS generation, including NADPH oxidase, further reducing oxidative burden at the source [22]. Moreover, sulfated polysaccharides from certain microalgae may exert antioxidant effects indirectly by modulating immune responses and attenuating inflammation-induced oxidative stress, possibly via cytokine signaling pathways and macrophage activation [23]. The antioxidant potential of microalgal compounds has been corroborated through a wide array of *in vitro*, *in vivo*, and limited clinical studies. For example, astaxanthin derived from *Haematococcus pluvialis* has demonstrated the ability to reduce lipid peroxidation and oxidative damage in various tissues, including the brain, liver, and muscle, in animal models [24]. Likewise, phycocyanin from *S. platensis* has shown promise in suppressing nitric oxide production and malondialdehyde (MDA) levels, thereby mitigating cellular oxidative injury [25]. In conclusion, the antioxidant activity of microalgal bioactives is mediated through a dual mechanism: direct chemical neutralization of reactive species, and activation of endogenous antioxidant defense systems. This dual action underlines their therapeutic promise in the prevention and management of diseases associated with oxidative stress and makes them valuable candidates for incorporation into functional foods, nutraceuticals, pharmaceuticals, and cosmeceutical formulations [26].

3.2 Antimicrobials and antivirals and antifungal properties

Microalgae are a diverse group of photosynthetic microorganisms with extraordinary metabolic plasticity, capable of producing a broad range of secondary metabolites. In recent years, the biomedical potential of microalgal bioactives—especially in combating pathogenic bacteria, viruses, and fungi—has gained considerable attention. The most prominent classes of bioactive compounds derived from microalgae and cyanobacteria that exhibit antimicrobial, antiviral, and antifungal activities, are reviewed, emphasizing both their biochemical mechanisms and pharmacological relevance [27].

3.2.1 Sulfated polysaccharides (SPs)

3.2.1.1 Structural features and sources

Sulfated polysaccharides are high-molecular-weight, anionic polymers excreted extracellularly by various microalgae—especially red algae (*Porphyridium*), *cyanobacteria,* and green microalgae. These polysaccharides contain repeating sugar units (e.g., xylose, rhamnose, uronic acids) modified by sulfate groups, which impart strong negative charges essential for their bioactivity [28].

SPs exhibit remarkable antiviral activity through competitive binding to viral glycoproteins such as gp120 (HIV-1), gD (HSV), or the spike (S) protein (SARS-CoV-2), thereby inhibiting viral adsorption or entry into host cells [29]. Their antibacterial effects are linked to interference with biofilm formation, membrane destabilization, or disruption of quorum sensing [30]. Antifungal activity—though less extensively studied—is hypothesized to occur *via* binding to chitin-rich fungal cell walls, impairing adhesion and growth [31].

Calcium spirulan (Ca-SP) from *S. platensis* exhibits potent inhibition against HIV-1, CMV, and HSV-1, with IC_{50} values in the low microgram range [32].

Carrageenans, especially from *red algae*, block HPV, HSV-2, DENV, and SARS-CoV-2 through direct viral envelope interference [33].

Fucoidan-like SPs, in genetically modified strains, demonstrate antiviral activity against influenza and SARS-related viruses [34].

3.2.2 Lectins

Lectins are carbohydrate-binding proteins capable of recognizing high-mannose and complex-type N-glycans on the surfaces of viruses, bacteria, and fungi. They are non-enzymatic yet highly selective, making them promising candidates for targeted antiviral strategies. Lectins exert broad-spectrum antiviral effects by binding to envelope glycoproteins, preventing viral entry or membrane fusion. Some lectins also exhibit antifungal and antibacterial properties, potentially through immunomodulation or interference with glycosylated receptors on microbial membranes [35].

Griffithsin, derived from red alga *Griffithsia*, binds to mannose-rich glycans of HIV-1, SARS-CoV-2, and Ebola with nanomolar affinity. Engineered variants like Q-GRFT have shown potent activity against *Candida albicans in vivo* [36].

Scytovirin, derived from the cyanobacterium *Scytonema varium*, demonstrates activity against Marburg and Ebola viruses, showing promise in preclinical antiviral platforms [32].

KAA-2 and *HRL40*, lectins from blue-green algae, have demonstrated inhibitory activity against influenza H3N2 strains [37].

3.2.3 Fatty acids and lipid-based derivatives

Microalgae produce a spectrum of polyunsaturated fatty acids (PUFAs), including eicosapentaenoic acid (EPA), docosahexaenoic acid (DHA), and specific monoglycerides, which display antimicrobial properties beyond their nutritional value. PUFAs exert antibacterial activity via membrane disruption, leading to increased permeability and loss of ionic balance in bacterial cells. Additionally, lipid derivatives have demonstrated antiviral effects by modulating host factors (e.g., inhibition of TMPRSS2 protease) critical for viral entry [38].

3.2.4 Pigments (carotenoids and phycobiliproteins)

Microalgal pigments such as astaxanthin, β-carotene, phycocyanin, and phycoerythrin are not only potent antioxidants but also exhibit notable antimicrobial and antiviral effects [39].

Phycocyanin from *Spirulina* inhibits the growth of *Pseudomonas aeruginosa* and *Staphylococcus aureus via* membrane disruption and ROS modulation [40].

Astaxanthin downregulates pro-inflammatory cytokines (TNF-α, IL-6) and enhances IL-10 production, thereby establishing an antiviral milieu and mitigating cytokine storms in viral infections like COVID-19 [41].

3.2.5 Peptides and non-ribosomal compounds

Several cyclic peptides, glycopeptides, and polyketide-like molecules have been isolated from microalgae and cyanobacteria with promising antimicrobial and antifungal activities.

Hassallidins are antifungal glycosylated lipopeptides that disrupt ergosterol-containing membranes in fungi [42].

Microviridins and *cyanobactins* inhibit serine proteases and bacterial quorum sensing, showing potential in addressing antimicrobial resistance (AMR) [43].

ECPI-2, isolated from *Chlorella* spp., is a cysteine protease inhibitor with potential for antiviral drug development due to its enzymatic targeting properties [44].

3.3 Anti-inflammatory and immunomodulatory agents

3.3.1 Anti-inflammatory properties

Among the most prominent bioactive compounds derived from microalgae is C-phycocyanin, a biliprotein abundantly present in species such as *S. platensis*. This compound exhibits potent antioxidant and anti-inflammatory properties by neutralizing ROS—including peroxyl, hydroxyl, and alkoxyl radicals—and by inhibiting lipid peroxidation. Mechanistically, C-phycocyanin exerts its anti-inflammatory effects through the upregulation of heme oxygenase-1 (HO-1) and suppression of the NF-κB signaling pathway, which collectively result in reduced expression of key pro-inflammatory cytokines such as TNF-α, IL-6, and IL-1β. In addition to phycocyanin, several other microalgal-derived compounds have demonstrated significant anti-inflammatory potential [45]:

Carotenoids, particularly xanthophylls and β-carotenes, play an important role in modulating inflammation:

Fucoxanthin, predominantly found in species such as *Phaeodactylum tricornutum* and *Tisochrysis lutea*, has been shown to inhibit key inflammatory signaling pathways including NF-κB, NLRP3 inflammasome, and JAK-STAT. *In vitro* studies reveal that fucoxanthin significantly downregulates LPS-induced secretion of IL-1β, IL-6, and TNF-α, and reduces activation of NLRP3 and cleaved caspase-1. In animal models of autoimmune arthritis, fucoxanthin supplementation not only mitigates joint inflammation but also rebalances gut microbiota and reduces systemic inflammatory markers [46].

Astaxanthin, a potent red carotenoid derived from *H. pluvialis*, exhibits robust anti-inflammatory activity. Experimental models of inflammatory bowel disease (IBD) have demonstrated that astaxanthin administration leads to marked suppression of TNF-α, IL-1β, and COX-2, largely through inhibition of NF-κB translocation. Furthermore, it has been noted for its antitumor and antimicrobial effects [47].

Lutein and zeaxanthin, two well-known xanthophylls, contribute to skin and ocular health by attenuating UV-induced oxidative stress. Their anti-inflammatory properties are mediated via downregulation of COX-2, inhibition of neutrophil migration, and enhancement of endogenous anti-inflammatory pathways [48].

Long-chain polyunsaturated fatty acids (LC-PUFAs), particularly EPA and DHA, derived from species like Nannochloropsis and Crypthecodinium, also exhibit notable anti-inflammatory efficacy. These omega-3 fatty acids suppress the expression of PGE_2, IL-8, and COX-2, and reduce neutrophilic infiltration and epidermal thickening in models of skin inflammation and burn injuries. These effects are thought to involve activation of PPARγ and inhibition of ERK/JNK signaling cascades [49].

Lin and colleagues [50] investigated the anti-inflammatory effects of a carotenoid extract derived from *Dunaliella salina*, containing α-carotene and β-carotene, on RAW264.7 macrophages infected with rabies-like virus. Their findings showed a concentration-dependent reduction of proinflammatory cytokines, including IL-1β, IL-6, TNF-α, and MCP-1. This effect was associated with a decrease in the activation of the NF-κB signaling pathway via Toll-Like Receptor 9 (TLR9), a process dependent on PI3K/Akt signaling.

3.3.2 Immunomodulatory effects

Microalgae are also a source of compounds capable of modulating immune responses. Among these, sulfated polysaccharides and sulfolipids stand out for their multifaceted regulatory roles [51]:

Fucoidan, a sulfated polysaccharide commonly extracted from *brown algae* such as *Fucus vesiculosus*, has gained significant attention for its immunomodulatory potential. This compound demonstrates high oral bioavailability, low toxicity, and the capacity to enhance immune surveillance by promoting phagocytosis, stimulating the release of regulatory cytokines, and engaging pattern recognition receptors such as TLRs and Dectin-1. Low-molecular-weight variants of fucoidan are currently under investigation in pharmacokinetic studies as potential therapeutic agents in clinical settings. For example, Ishikawa et al. showed that fucoxanthin and its metabolite fucoxanthinol significantly inhibited the viability of adult T-cell leukemia (ATL) cells, a malignancy caused by human T-cell leukemia virus type 1 (HTLV-1) [52].

In addition, low molecular weight metabolites such as apigenin and catechin, identified through *in silico* screening in certain microalgal species, have shown promise in modulating inflammatory receptors and enhancing antiviral and immune-regulating responses [53].

3.4 Anticancer compounds

Microalgae represent a prolific source of bioactive compounds with demonstrated anticancer properties, many of which have been characterized by unique chemical structures and selective cytotoxicity. These natural products interfere with multiple cancer hallmarks, including cell proliferation, apoptosis evasion, angiogenesis, and metastasis, offering promising alternatives or adjuncts to conventional chemotherapies [54].

3.4.1 C-Phycocyanin

C-Phycocyanin (C-PC), a phycobiliprotein abundantly found in *S. platensis*, has been extensively studied for its potent anticancer activity. Mechanistically, C-PC induces cell cycle arrest at the G0/G1 phase and promotes apoptosis *via* upregulation of intracellular ROS and downregulation of antiapoptotic proteins such as Bcl-2. The activation of caspase cascades further facilitates apoptotic progression. *In vitro* and *in vivo* studies have shown that C-PC can synergistically enhance the efficacy of chemotherapeutic agents [55]. For instance, the co-administration of C-PC with low-dose topotecan in prostate cancer cells (LNCaP) significantly amplified apoptotic responses compared to monotherapy [56]. Furthermore, C-PC has been reported to reduce metastasis in colorectal cancer mouse models by downregulating pro-angiogenic and invasion-related markers such as VEGF-A, MMP-2, MMP-9, and HIF-1α, alongside modulating chemokines, including MCP-1 and MIP-1 [57].

3.4.2 Carotenoids and related metabolites

Among microalgal carotenoids, astaxanthin from *H. pluvialis* has shown marked antiproliferative activity in gastric cancer cell lines (KATO-III, SNU-1), primarily through G0/G1 cell cycle arrest, increased p27 expression, and inhibition of ERK signaling pathways [58]. Another noteworthy class involves polyunsaturated aldehydes (PUAs), produced by diatoms such as *Skeletonema marinoi* and *Thalassiosira rotula*. These compounds exert strong antiproliferative effects on colon (Caco-2) and lung (A549) cancer cells, although their precise cellular targets remain under investigation [59].

3.4.3 Cyanobacterial metabolites and antimetabolites

Cyanobacteria also contribute to structurally diverse and biologically active metabolites. *Cryptophycins,* isolated from *Nostoc* species, are potent depsipeptides that disrupt microtubule dynamics, akin to vinblastine, inducing mitotic arrest and apoptosis. Their nanomolar-range cytotoxicity underscores their pharmaceutical potential [60]. Calothrixins A and B, marine-derived indole alkaloids, have demonstrated inhibitory effects against colon cancer cells (e.g., HCT-116) at sub-micromolar concentrations [61]. Similarly, malyngamides—particularly Malyngamide C and its 8-epi isomer, extracted from *Lyngbya majuscula*—have shown significant cytotoxicity against HT-29 cells. Additional *cyanobacterial* compounds such as Merocyclophanes A/B and Hierridin B also exhibit strong antitumor activity, likely mediated by modulation of oxidative stress and interference with oncogenic signaling cascades [62].

Euglenophycin, a piperidine derivative from *Euglena sanguinea*, has emerged as a promising antineoplastic compound. Preliminary studies suggest it acts through the inhibition of the PI3K/AKT/mTOR axis and suppression of VEGF-mediated angiogenesis, while promoting G1 cell cycle arrest [63].

3.4.4 Sulfated exopolysaccharides (EPS)

Sulfated polysaccharides, particularly those secreted by red microalgae such as *Porphyridium aerugineum*, exhibit dose-dependent cytotoxicity in cancer cell lines [64]. These EPSs have demonstrated selective inhibition of breast (MCF-7) and colon (HCT-8) cancer cells, with IC_{50} values ranging between ~800 to 1100 μg/mL. Their anticancer activity is closely associated with molecular characteristics such as degree of sulfation, uronic acid content, and molecular weight [65].

3.5 Neuroprotective and cardioprotective agents

Microalgae are increasingly recognized as rich reservoirs of bioactive compounds with significant therapeutic potential. Among these, a growing body of research highlights their dual neuroprotective and cardioprotective effects, which are mainly attributed to their anti-inflammatory, antioxidant, and mitochondrial-stabilizing activities. Such properties are particularly relevant in the context of neurodegenerative and cardiovascular diseases, which share common pathogenic pathways, including oxidative stress, inflammation, mitochondrial dysfunction, and apoptosis [66].

3.5.1 Astaxanthin

Astaxanthin, a red-pigmented xanthophyll predominantly derived from *H. pluvialis*, is considered one of the most potent natural antioxidants [67]. Its lipophilic structure enables it to readily cross the blood–brain barrier, making it particularly valuable for central nervous system (CNS) protection. Experimental studies have demonstrated that astaxanthin significantly attenuates oxidative stress by scavenging ROS, upregulating endogenous antioxidant enzymes (such as SOD, CAT, and GSH), and suppressing pro-inflammatory signaling pathways, including NF-κB and MAPK. In neurodegenerative models such as ischemic stroke or Alzheimer's disease, astaxanthin has shown the capacity to preserve neuronal viability, reduce neuroinflammation, and improve cognitive function [68]. Likewise, in cardiovascular models, it reduces lipid peroxidation, inhibits cardiomyocyte apoptosis, and lowers circulating biomarkers of myocardial damage such as CK-MB and LDH. Its cardioprotective role has also been observed in doxorubicin-induced cardiotoxicity and myocardial ischemia-reperfusion injury, where it stabilizes mitochondrial membranes and preserves cardiac contractility [69].

3.5.2 Fucoxanthin

Fucoxanthin, a carotenoid found mainly in diatoms such as *P. tricornutum*, has attracted attention for its pleiotropic health effects. It exerts potent anti-inflammatory actions by modulating the Nrf2 pathway and inhibiting the NLRP3 inflammasome complex. In preclinical stroke models, fucoxanthin has demonstrated neuroprotective effects by enhancing blood–brain barrier integrity, reducing infarct volume, and suppressing proinflammatory cytokines (e.g., TNF-α, IL-1β) [70]. Its cardioprotective potential is equally compelling. Fucoxanthin has been reported to lower blood pressure, improve endothelial function, and prevent cardiac apoptosis by modulating key signaling cascades such as PI3K/AKT and ERK. It also contributes to mitochondrial homeostasis, which is critical in preventing cardiac hypertrophy and ischemia-induced injury [71].

3.5.3 Omega-3 fatty acids (EPA and DHA)

Microalgae, such as *Schizochytrium spp.* and *Nannochloropsis spp.*, are well-established sustainable sources of long-chain omega-3 polyunsaturated fatty acids (PUFAs), particularly EPA and DHA. These fatty acids have been extensively studied for their cardiovascular and neurological benefits. Mechanistically, EPA and

DHA modulate membrane fluidity, downregulate inflammatory cytokines (e.g., IL-6, TNF-α, CRP), and activate nuclear receptors such as PPARγ. This leads to the inhibition of pro-inflammatory MAPK and JNK signaling pathways, ultimately reducing chronic inflammation in both brain and heart tissues [72]. Furthermore, EPA and DHA enhance neuronal membrane stability and synaptic plasticity while reducing amyloid-beta accumulation, a hallmark of Alzheimer's disease [73]. In cardiovascular systems, they improve lipid profiles by increasing HDL and reducing LDL and triglycerides, thereby reducing atherosclerosis risk and myocardial infarction.

4. Mechanisms of action

This section explores the molecular mechanisms through which microalgal bioactive compounds exert their therapeutic effects. Drawing on peer-reviewed scientific literature, it focuses on key cellular signaling pathways modulated by these compounds, with an emphasis on their antioxidant, anti-inflammatory, antiproliferative, and neuroprotective activities.

4.1 Molecular pathways targeted by microalgal compounds

4.1.1 PI3K/AKT/mTOR pathway

Several microalgal compounds, including phycocyanin and euglenophycin, have been shown to interfere with the PI3K/AKT/mTOR pathway—a central signaling axis regulating cell growth, survival, and autophagy. Phycocyanin suppresses AKT phosphorylation and downstream effectors such as p70S6K, thereby inducing cell cycle arrest (G0/G1 phase), apoptosis, and autophagic processes. Euglenophycin, a polyketide compound structurally similar to solenopsin, similarly disrupts this pathway, leading to reduced tumor cell viability and inhibition of angiogenesis by downregulating VEGF expression [74].

4.1.2 NF-κB and Nrf2 pathways

Phycobiliproteins and carotenoids, such as astaxanthin and fucoxanthin, modulate oxidative stress and inflammatory signaling by regulating NF-κB and Nrf2 pathways. These compounds downregulate pro-inflammatory mediators, such as COX-2, cyclin D1, and Bcl-2, while simultaneously promoting the expression of pro-apoptotic proteins including Bax and caspase-3. Moreover, activation of the Nrf2 pathway enhances the cellular antioxidant response *via* upregulation of key detoxifying enzymes, such as SOD, CAT, GPx, and the glutathione system [75, 76].

4.1.3 MAPK/ERK and p38MAPK pathways

Carotenoids and polyphenolic compounds derived from microalgae also influence the MAPK/ERK and p38MAPK signaling cascades. These pathways play pivotal roles in cell proliferation, migration, and differentiation. Fucoxanthin, for example, reduces the phosphorylation of ERK1/2 and p38, thereby inhibiting tumor cell migration and promoting ROS-mediated apoptosis. Additionally, suppression of CDK4 and cyclin D1 expression has been observed, supporting the anti-proliferative profile of these compounds [77].

4.1.4 JAK/STAT pathway

Although direct evidence of JAK/STAT modulation by microalgal metabolites is still emerging, parallels with other marine bioactives suggest a similar regulatory potential. Suppression of STAT3 phosphorylation—a key driver of inflammation and oncogenesis—has been documented in related compounds, implying that certain microalgal extracts may exert comparable immunomodulatory and antiproliferative effects [78].

4.1.5 AMPK, Sirtuins, and autophagy

Bioactives, such as long-chain polyunsaturated fatty acids (PUFAs), β-1,3-glucans, and sulfated polysaccharides from microalgae, can activate the AMPK pathway, promoting cellular energy homeostasis and mitochondrial biogenesis via PGC-1α upregulation. Concurrently, these compounds may enhance SIRT1 activity, contributing to improved cellular resilience and induction of protective autophagy. Together, these mechanisms support metabolic regulation and cytoprotection in stress conditions [22].

4.1.6 UV defense and antioxidant mechanisms

Mycosporine-like amino acids (MAAs), produced by various microalgae, exhibit potent UV-absorbing and radical-scavenging properties. These small-molecule metabolites reduce oxidative DNA damage and contribute to photoprotection by stabilizing cellular redox balance under environmental stressors [79].

4.2 Pharmacokinetics and bioavailability challenges

Despite the promising therapeutic potential of microalgal bioactives—particularly phycocyanin (a water-soluble phycobiliprotein) and fucoxanthin (a lipid-soluble carotenoid)—their clinical translation remains hindered by significant pharmacokinetic and bioavailability limitations [80]. Here, we examine in detail the absorption, metabolism, chemical stability, tissue distribution, and delivery strategies based on current scientific evidence.

4.2.1 Absorption and early bioavailability

Fucoxanthin (FX) displays extremely limited oral bioavailability in humans: following administration of a 31 mg dose *via* kombu extract, FX itself was undetectable in plasma. Its principal metabolite, fucoxanthinol (FXol), reached a Cmax of approximately 44.2 nmol/L, at a Tmax of ~4 hours, with a terminal half-life of ~7 hours and an AUC∞ of ~663.7 nmol·h/L [81]. Amarouciaxanthin A—a secondary hepatic metabolite—was not detected in plasma. Preclinical murine studies show that FX is rapidly hydrolyzed to FXol, which accumulates in tissues such as liver, lungs, and adipose tissue within 1-hour post-dose, and remains detectable for up to 72 hours. In contrast, phycocyanin (C-PC) suffers from negligible oral absorption due to its high molecular weight and sensitivity to acidic and enzymatic degradation in the gastrointestinal tract. Without encapsulation or structural protection, it is entirely degraded before systemic exposure [82].

4.2.2 Metabolism and tissue distribution

In mice, FXol exhibits highest tissue accumulation in the liver (AUC∞ ≈ 4680 nmol·h/g), while amarouciaxanthin A preferentially accumulates in adipose tissues (AUC∞ ≈ 4630 nmol·h/g). Notably, human PK profiles show dramatically lower exposure, with FXol Cmax and AUC values representing only ~33% and ~ 46% of those observed in mice—despite lower per-kilogram dosage in humans. These interspecies discrepancies underscore the challenges of extrapolating preclinical data to humans [83].

4.2.3 Chemical stability and storage parameters

Fucoxanthin is highly prone to oxidative degradation, thermal isomerization, and photolysis, with measured degradation occurring within days to weeks at ambient temperature in the absence of stabilizing excipients. Similarly, phycocyanin is chemically unstable under elevated temperature (>40°C), improper pH, and ionic conditions. Empirical studies indicate that maintaining >90% of its activity over several weeks requires storage at pH 6–7, ≤4°C, and the use of stabilizing sugars or polymers [84].

4.2.4 Formulation strategies to enhance bioavailability

Multiple advanced delivery techniques have been applied to overcome these hurdles:

Fucoxanthin: Use of nano emulsions, cyclodextrin or maltodextrin microencapsulation, and coacervation-based lipid carriers significantly increased its bioaccessibility in animal models. These systems improved chemical stability, promoted sustained release, and enhanced FXol absorption several-fold compared to unformulated FX [85, 86].

Phycocyanin: Encapsulation within alginate beads, liposomes, protein–sugar matrices, or nanocarriers effectively shields it from gastric degradation and enhances its release in intestinal conditions, with improved antioxidant retention *in vitro* [87].

5. Challenges and limitations

5.1 Standardization and quality control

One of the major challenges in the development and commercialization of microalgae-derived therapeutic and nutraceutical products lies in the lack of comprehensive standardization frameworks and robust quality control systems. The bioactive compounds extracted from microalgae are often structurally complex, consisting of a heterogeneous matrix of primary and secondary metabolites such as polysaccharides, proteins, polyunsaturated fatty acids, pigments, and other functional molecules. These compositions are highly sensitive to environmental and cultivation parameters. Even minor fluctuations in factors such as temperature, light intensity, nutrient composition, pH, and aeration can lead to substantial variations in both the qualitative and quantitative profiles of the final product. As a result, achieving consistent and predictable production on an industrial scale requires stringent control over every step of the process—from cultivation and harvesting to extraction and formulation. Adding to the complexity is the absence of internationally harmonized analytical standards for characterizing, quantifying, and validating the purity of microalgal products. Unlike conventional pharmaceuticals, which are regulated by well-established pharmacopeias such as the USP, EMA, or WHO guidelines, many microalgae-derived compounds still lack validated methodologies for quality assessment. This gap hinders traceability, reproducibility, and ultimately, regulatory approval, posing a significant barrier to global market access. In this context, the implementation of Quality by Design (QbD) approaches can be instrumental. QbD emphasizes the identification and control of Critical Quality Attributes (CQAs) and Critical Process Parameters (CPPs) throughout the production cycle. For example, parameters such as phycocyanin content, omega-3 fatty acid concentration, or microbial purity can be considered CQAs, while cultivation time, nutrient feed rates, and photoperiod conditions may function as CPPs. The goal is to define a robust design space wherein acceptable product quality is maintained despite variations within predefined limits. Furthermore, while analytical tools, such as HPLC, GC-MS, LC-MS/MS, NMR, and FTIR, offer powerful capabilities for real-time quality monitoring, their effectiveness is limited by the lack of validated reference standards for microalgal matrices. Many producers also face challenges in consistently evaluating and mitigating potential contaminants such as microbial impurities, heavy metals, solvent residues, and endotoxins, particularly in large-scale operations.

In summary, the absence of standardized protocols for quality assurance not only increases the risk of product inconsistency but also significantly impedes the international competitiveness of microalgal technologies. Addressing these limitations will require coordinated efforts toward the development of global standards specific to microalgae-derived products, the validation of analytical techniques, and the integration of QbD and GMP frameworks into the manufacturing pipeline. Only through such systemic improvements can the field ensure the production of safe, effective, and reproducible microalgal therapeutics and supplements [88].

5.2 Toxicological concerns

While microalgae hold tremendous promise as sources of therapeutic and nutritional compounds, ensuring their toxicological safety remains a critical component of product development. The biological complexity of these organisms—combined with the diversity of environmental and cultivation conditions—means that some species may produce toxic secondary metabolites under stress, such as extreme light, nutrient deprivation, or changes in temperature and pH. Compounds like microcystins, anatoxins, saxitoxins, or β-N-methylamino-L-alanine (BMAA) have been identified in certain strains, especially among cyanobacteria, and can pose serious risks to human health and the environment if not properly monitored.

Rigorous toxicological testing—both *in vitro* (cell-based assays) and *in vivo* (animal models)—is essential to evaluate the safety profile of microalgal products. These studies are governed by internationally recognized guidelines, including those issued by the OECD (Organization for Economic Co-operation and Development), ICH (International Council for Harmonization of Technical Requirements for Pharmaceuticals), and the World Health Organization (WHO). Key areas of assessment include:

- *Acute and chronic toxicity* to determine immediate and long-term effects of exposure.
- *Genotoxicity* to evaluate the potential for DNA damage, mutagenesis, or chromosomal abnormalities.
- *Allergenicity* to assess immune responses or hypersensitivity reactions in consumers.
- *Pharmacokinetics (ADME)* to understand how compounds are absorbed, distributed, metabolized, and excreted in biological systems.

Despite the long history of use and GRAS (Generally Recognized as Safe) status for species like *Spirulina* and *Chlorella* in many countries, mild adverse reactions—such as gastrointestinal discomfort, skin irritation, or allergic symptoms—have occasionally been reported, particularly in sensitive individuals. These instances highlight the importance of thorough safety evaluations, even for widely consumed strains. A significant limitation in the field remains the lack of comprehensive toxicological data for many novel or lesser-known microalgal species. Numerous compounds identified from these organisms have yet to undergo preclinical evaluation, creating a bottleneck in regulatory approval and market entry. Addressing this gap requires well-designed toxicology studies, spanning different doses, exposure routes (oral, topical, injectable), and time frames to capture potential risks accurately. Only through robust, transparent, and standardized toxicological assessment can the therapeutic and nutritional potential of microalgae be fully realized—without compromising consumer safety or environmental integrity [89].

6. Future directions

6.1 Genomic and metabolomic approaches

Recent advances in systems biotechnology have brought genomic, transcriptomic, proteomic, and metabolomic tools to the forefront of microalgal research, opening new possibilities for the discovery and optimization of therapeutically relevant compounds. These integrative approaches allow researchers to unravel the complex biological networks underpinning metabolite production, enabling rational strain engineering and fine-tuned cultivation strategies for improved yields of bioactive molecules [90].

In the realm of *genomics*, next-generation sequencing (NGS) has facilitated the full genome annotation of several medicinally important microalgae, including *Chlorella vulgaris*, *Nannochloropsis oceanica*, *H. pluvialis*, and *P. tricornutum*. This genomic insight enables the identification of key biosynthetic genes and regulatory pathways responsible for the production of high-value compounds such as astaxanthin, EPA, DHA, sulfated polysaccharides, and various antioxidants. For instance, recent studies have identified over 50 genes involved in astaxanthin biosynthesis in *H. pluvialis*, highlighting regulatory responses to light intensity and temperature fluctuations [91, 92]. The rise of genome editing tools, particularly CRISPR-Cas9, TALENs, and zinc finger nucleases, has further accelerated functional genomics in microalgae. These platforms enable precise modifications of metabolic pathways—either by knocking out competing routes or upregulating desired enzymatic steps—thus enhancing the overall flux toward therapeutic target molecules [93]. Genetic circuits and custom plasmid systems designed for microalgal hosts are becoming increasingly sophisticated, facilitating synthetic biology applications for strain optimization.

Transcriptomic profiling, particularly *via* RNA-Seq, provides an essential layer of regulation by capturing dynamic gene expression in response to environmental and chemical stimuli. When combined with *proteomic* data, this offers a multi-tiered understanding of metabolic regulation, especially under stress conditions like nitrogen limitation, high salinity, or light-induced oxidative pressure. These omics platforms have revealed the upregulation of lipid biosynthetic enzymes and carotenoid-associated genes under nutrient stress, aligning gene expression patterns with metabolite accumulation profiles [94]. Perhaps the most immediate insights into cellular metabolic performance come from *metabolomics*. Techniques such as GC-MS, LC-MS/MS, and NMR spectroscopy enable the detailed characterization of intracellular and extracellular metabolites, including fatty acids, sterols, polysaccharides, peptides, and other pharmacologically active compounds. These profiles help identify metabolic bottlenecks, optimize induction conditions, and support downstream bioprocessing design [95, 96].

Importantly, integrating these multi-omics datasets within a systems biology framework allows for the construction of comprehensive metabolic and regulatory models. These *in silico* tools can simulate cellular responses under varying cultivation regimes and guide the rational design of microalgal strains for targeted compound synthesis. For example, such modeling has supported the development of EPA-overproducing strains of *Nannochloropsis* for industrial-scale production under photobioreactor conditions [97].

In conclusion, the convergence of omics-based technologies with advanced analytics and machine learning is set to revolutionize therapeutic microalgae biotechnology. As data repositories grow and computational tools improve, a new generation of designer strains with enhanced resilience, metabolic productivity, and economic viability is likely to emerge—bringing microalgae-derived therapeutics closer to mainstream clinical and commercial applications.

6.2 Synthetic biology and genetic engineering

In recent years, *synthetic biology and genetic engineering* have emerged as pivotal tools for the targeted modification of microalgae, especially in the context of novel therapeutic applications. Unlike traditional methods of strain improvement—which rely on natural selection or random mutagenesis—modern genetic engineering technologies now allow for precise design and regulation of metabolic pathways within these organisms. These engineered strategies can significantly enhance the biosynthetic capacity of microalgae to produce high-value bioactive compounds, such as antioxidants, polysaccharides, alkaloids, and even therapeutic recombinant proteins [98]. One of the central goals of synthetic biology in microalgal systems is metabolic pathway optimization, aimed at the efficient production of specific target molecules. For instance, using tools such as CRISPR/Cas9, RNA interference (RNAi), and the design of synthetic or inducible promoters, researchers have successfully fine-tuned the biosynthesis of carotenoids or sulfated polysaccharides in species such as *Chlamydomonas reinhardtii* and *Phaeodactylum tricornutum*. Synthetic biology also enables the in-cell biosynthesis of non-native products, effectively turning microalgae into biofactories for the production of novel compounds, including vaccines, therapeutic enzymes, and monoclonal antibodies. These advances not only improve the efficiency of biological production systems but also contribute to enhanced biosafety and process control, particularly in ensuring product purity and consistency [99]. From a pharmaceutical biotechnology perspective, one of the ongoing challenges is the stable and high-level expression of heterologous genes in microalgal hosts. Recent developments have focused on the construction of species-specific vectors, optimization of endogenous gene regulatory elements, and synthetic operon engineering to overcome this bottleneck. Moreover, the integration of computational tools and multiscale modeling allows for more precise simulation and reconstruction of complex metabolic networks [100].

Ultimately, the convergence of synthetic biology with other "omics" disciplines—such as genomics, transcriptomics, and metabolomics—is shaping a new systems-level approach to designing the next generation of algal therapeutics. This interdisciplinary synergy not only accelerates product development but also enhances the quality, efficacy, and scalability of microalgae-derived medical treatments.

7. Conclusion

Microalgae represent a promising yet underexplored reservoir of bioactive compounds with significant implications for modern medicine. Over the past two decades, scientific investigations have increasingly highlighted the diverse pharmacological properties of microalgal metabolites, including carotenoids, phycobiliproteins, polyunsaturated fatty acids, polysaccharides, and various phenolic and alkaloid compounds. These substances have demonstrated a broad spectrum of therapeutic effects, such as antioxidant, anti-inflammatory, antimicrobial, anticancer, immunomodulatory, antiviral, and neuroprotective activities.

Species like *Spirulina* and *Chlorella* have shown encouraging results in clinical and preclinical studies related to metabolic disorders, chronic inflammation, cardiovascular health, and immune function. Notably, some of these microalgae-derived compounds can influence key molecular pathways, modulate oxidative stress responses, and regulate immune cell activity, suggesting their potential role in both preventive and adjunctive therapies. The diversity and specificity of these bioactive molecules make microalgae a unique platform for the development of natural therapeutics.

With the advancement of biotechnological tools, such as omics technologies, genetic engineering, and controlled bioreactor cultivation, the potential for integrating microalgae into pharmaceutical and clinical frameworks is becoming increasingly feasible. Current research is moving beyond basic biological activity, aiming to harness microalgae for more sophisticated applications—such as targeted drug delivery systems, recombinant protein production, vaccine development, and even in personalized medicine platforms. Nevertheless, several challenges remain before full-scale medical adoption can occur. These include regulatory complexities, the need for standardized cultivation and extraction protocols, comprehensive safety evaluations, and scalable production systems. Furthermore, human clinical trials are still limited in scope and number, and robust data on pharmacokinetics, bioavailability, and long-term effects are essential for regulatory approval.

Despite these hurdles, the environmental sustainability of microalgae cultivation, coupled with growing demand for plant-based and natural therapeutics, places them in an increasingly favorable position. As interdisciplinary collaboration between pharmacology, biotechnology, and environmental science continues to expand, the incorporation of microalgae into the medical and pharmaceutical landscape appears not only promising but inevitable. Their unique combination of therapeutic potential, renewable sourcing, and molecular diversity positions them as a cornerstone of future biomedicine.

Author details

Negar Nasri[1*] and Homeira Zare Chavoshy[2]

1 Department of Life Science Engineering, Faculty of New Sciences and Technologies, University of Tehran, Isfahan, Iran

2 Department of Cellular and Molecular, Shahrekord Branch, Islamic Azad University, Iran

*Address all correspondence to: negar.nasry@gmail.com

References

[1] Queiroz MI, Vieira JG, Maroneze MM. Morphophysiological, structural, and metabolic aspects of microalgae. In: Handbook of Microalgae-Based Processes and Products. Academic Press; 2020. pp. 25-48

[2] Thoré ES et al. Microalgae. Current Biology. 2023;**33**(3):R91-R95

[3] Borowitzka MA. Microalgae in medicine and human health: A historical perspective. In: Microalgae in Health and Disease Prevention. Academic Press; 2018. pp. 195-210

[4] Wu Q et al. The antioxidant, immunomodulatory, and anti-inflammatory activities of spirulina: An overview. Archives of Toxicology. 2016;**90**(8):1817-1840

[5] Galasso C et al. Microalgal derivatives as potential nutraceutical and food supplements for human health: A focus on cancer prevention and interception. Nutrients. 2019;**11**(6):1226

[6] Singh J, Saxena RC. An introduction to microalgae: Diversity and significance. In: Handbook of Marine Microalgae. Academic Press; 2015. pp. 11-24

[7] Metting F Jr. Biodiversity and application of microalgae. Journal of Industrial Microbiology. 1996;**17**(5):477-489

[8] Jassim YA, Awadh EFA, Al-Amery SMH. A review of general properties of blue-green algae (cyanobacteria). Biomedicine and Chemical Sciences. 2023;**2**(2):143-148

[9] Park NY, Kim B, Kim WY, Lim AS. Understanding the biological rhythms of eukaryotic microalgae and their biotechnological and ecological applications. Journal Of Plant Biology. 2025;**68**:1-18, 155-172

[10] Leliaert F. Diversity and Evolution of Algae: Primary Endosymbiosis. Dordrecht, Netherlands: Springer; 2015

[11] Friedl T. The evolution of the green algae. In: Origins of Algae and their Plastids. Vienna: Springer Vienna; 1997. pp. 87-101

[12] Barsanti L, Gualtieri P. Algae: Anatomy, Biochemistry, and Biotechnology. 3rd ed. CRC press; 2022

[13] Kuppusamy P et al. Potential pharmaceutical and biomedical applications of diatoms microalgae-an overview. Indian Journal of Geo-Marine Sciences. 2017;**46**:663-667

[14] Serôdio J, Lavaud J. Diatoms and their ecological importance. In: Life below Water. Cham: Springer International; 2022. pp. 304-312

[15] Spector DL editor. Dinoflagellates. Academic Press; 2012

[16] Hoppenrath M. Dinoflagellate taxonomy—A review and proposal of a revised classification. Marine Biodiversity. 2017;**47**(2):381-403

[17] Nicholls KH. Haptophyte algae. In: Freshwater Algae of North America. Academic Press; 2015. pp. 587-605

[18] Bicudo CEDM, Menezes M. Phylogeny and classification of Euglenophyceae: A brief review. Frontiers in Ecology and Evolution. 2016;**4**:17

[19] Coulombier N, Jauffrais T, Lebouvier N. Antioxidant compounds

from microalgae: A review. Marine Drugs. 2021;**19**(10):549

[20] Goiris K et al. Antioxidant potential of microalgae in relation to their phenolic and carotenoid content. Journal of Applied Phycology. 2012;**24**(6):1477-1486

[21] Melloni M et al. Microalgae as a nutraceutical tool to antagonize the impairment of redox status induced by SNPs: Implications on insulin resistance. Biology. 2023;**12**(3):449

[22] Vignaud J et al. Microalgae produce antioxidant molecules with potential preventive effects on mitochondrial functions and skeletal muscular oxidative stress. Antioxidants. 2023;**12**(5):1050

[23] Lakhrem M et al. Anti-vasculogenic, antioxidant, and anti-inflammatory activities of sulfated polysaccharide derived from *Codium tomentosum*: Pharmacokinetic assay. Pharmaceuticals. 2024;**17**(6):672

[24] Oslan SNH et al. *Haematococcus pluvialis* as a potential source of astaxanthin with diverse applications in industrial sectors: Current research and future directions. Molecules. 2021;**26**(21):6470

[25] Citi V et al. Nutraceutical features of the phycobiliprotein C-phycocyanin: Evidence from *Arthrospira platensis* (spirulina). Nutrients. 2024;**16**(11):1752

[26] Yang N et al. Study on bioactive compounds of microalgae as antioxidants in a bibliometric analysis and visualization perspective. Frontiers in Plant Science. 2023;**14**:1144326

[27] Ognistaia A, Markina ZV, Orlova TY. Antimicrobial activity of marine microalgae. Russian Journal of Marine Biology. 2022;**48**(4):217-230

[28] Severo IA et al. Microalgae-derived polysaccharides: Potential building blocks for biomedical applications. World Journal of Microbiology and Biotechnology. 2022;**38**(9):150

[29] Kumar A et al. Algal metabolites can be an immune booster against COVID-19 pandemic. Antioxidants. 2022;**11**(3):452

[30] Behzadnia A, Moosavi-Nasab M, Oliyaei N. Anti-biofilm activity of marine algae-derived bioactive compounds. Frontiers in Microbiology. 2024;**15**:1270174

[31] Schmid B et al. Antifungal properties of aqueous microalgal extracts. Bioresource Technology Reports. 2022;**18**:101096

[32] Prabhu S, Vijayakumar S, Praseetha P. Cyanobacterial metabolites as novel drug candidates in corona viral therapies: A review. Chronic Diseases and Translational Medicine. 2022;**8**(03):172-183

[33] Oliyaei N, Moosavi-Nasab M, Mazloomi SM. Therapeutic activity of fucoidan and carrageenan as marine algal polysaccharides against viruses. 3 Biotech. 2022;**12**(7):154

[34] Akl FM et al. Antimicrobial, antidiabetic, antiviral, and antioxidant activities of fucoidan extracted from the brown seaweed Padina pavonica. BMC Biotechnology. 2025;**25**(1):70

[35] Ashraf A et al. Microalgae-derived peptides: Exploring bioactivities and functional food innovations. Journal of Agricultural and Food Chemistry. 2025;**73**(2):1000-1013

[36] Fernández Romero JA et al. Algal and cyanobacterial lectins and their antimicrobial properties. Marine Drugs. 2021;**19**(12):687

[37] Nazmul T et al. Capture and neutralization of SARS-CoV-2 and influenza virus by algae-derived lectins with high-mannose and core fucose specificities. Microbiology and Immunology. 2023;**67**(7):334-344

[38] Cepas V et al. Microalgae and cyanobacteria strains as producers of lipids with antibacterial and antibiofilm activity. Marine Drugs. 2021;**19**(12):675

[39] Lortou U, Gkelis S. Antibacterial activity, pigments, and biomass content of microalgae isolated from Greece. Journal of Biological Research-Thessaloniki. 2023;**30**

[40] Ilieva Y et al. Antimicrobial activity of Arthrospira (former spirulina) and Dunaliella related to recognized antimicrobial bioactive compounds. International Journal of Molecular Sciences. 2024;**25**(10):5548

[41] Wu Y et al. Astaxanthin targets IL-6 and alleviates the LPS-induced adverse inflammatory response of macrophages. Food & Function. 2024;**15**(8):4207-4222

[42] Fewer DP et al. Chemical diversity and cellular effects of antifungal cyclic lipopeptides from cyanobacteria. Physiologia Plantarum. 2021;**173**(2):639-650

[43] do Amaral SC et al. Current knowledge on microviridin from cyanobacteria. Marine Drugs. 2021;**19**(1):17

[44] Ishihara M et al. Purification and characterization of extracellular cysteine protease inhibitor, ECPI-2, from chlorella sp. Journal of Bioscience and Bioengineering. 2006;**101**(2):166-171

[45] Piovan A et al. The effect of C-phycocyanin on microglia activation is mediated by toll-like receptor 4. International Journal of Molecular Sciences. 2022;**23**(3):1440

[46] Lee A-H et al. Fucoxanthin from microalgae *Phaeodactylum tricornutum* inhibits pro-inflammatory cytokines by regulating both NF-κB and NLRP3 inflammasome activation. Scientific Reports. 2021;**11**(1):543

[47] Zhang H et al. Protective effects of Astaxanthin against oxidative stress: Attenuation of TNF-α-induced oxidative damage in SW480 cells and Azoxymethane/dextran sulfate sodium-induced colitis-associated cancer in C57BL/6 mice. Marine Drugs. 2024;**22**(10):469

[48] Kim E-A et al. Antioxidant, antiviral, and anti-inflammatory activities of lutein-enriched extract of Tetraselmis species. Marine Drugs. 2023;**21**(7):369

[49] Mahata C. Extraction and Enrichment of Fatty Acids from Marine Microalgae. Marine Molecules from Algae and Cyanobacteria journal, science direct. 2025;**2025**:41-57

[50] Lin HW, Chen YC, Liu CW, Yang DJ, Chen SY, Chang TJ, et al. Regulation of virus-induced inflammatory response by Dunaliella salina alga extract in macrophages. Food and Chemical Toxicology. 2014;**71**:159-165

[51] Riccio G, Lauritano C. Microalgae with immunomodulatory activities. Marine Drugs. 2019;**18**(1):2

[52] Apostolova E et al. Immunomodulatory and anti-inflammatory effects of fucoidan: A review. Polymers. 2020;**12**(10):2338

[53] Mishra J, Sahu A, Kushwaha N, Soni A, Singh S, Dorwal D, et al. Nutraceuticals as immunomodulators. In: The Nature of

Nutraceuticals. Apple Academic Press; 2025. pp. 207-251

[54] Talero E et al. Bioactive compounds isolated from microalgae in chronic inflammation and cancer. Marine Drugs. 2015;**13**(10):6152-6209

[55] Bannu SM et al. Potential therapeutic applications of C-phycocyanin. Current Drug Metabolism. 2019;**20**(12):967-976

[56] Gantar M, Dhandayuthapani S, Rathinavelu A. Phycocyanin induces apoptosis and enhances the effect of topotecan on prostate cell line LNCaP. Journal of Medicinal Food. 2012;**15**(12):1091-1095

[57] Kefayat A et al. C-phycocyanin: A natural product with radiosensitizing property for enhancement of colon cancer radiation therapy efficacy through inhibition of COX-2 expression. Scientific Reports. 2019;**9**(1):19161

[58] Ranga, Rao A. Production of Astaxanthin from Cultured Green Alga Haematococcus Pluvialis and its Biological Activities. Doctoral dissertation, University of Mysore; 2011

[59] Sansone C et al. Diatom-derived polyunsaturated aldehydes activate cell death in human cancer cell lines but not normal cells. PLoS One. 2014;**9**(7):e101220

[60] Lee H, Nihan K, Kwon YR. Cyanobacterial peptides in anticancer therapy: A comprehensive review of mechanisms, clinical advances, and biotechnological innovation. Marine Drugs. 2025;**23**(6):233

[61] Matulja D et al. Marine natural products with high anticancer activities. Current Medicinal Chemistry. 2020;**27**(8):1243-1307

[62] Cock IE, Cheesman MJ. A review of the antimicrobial properties of cyanobacterial natural products. Molecules. 2023;**28**(20):7127

[63] Aldholmi M et al. Euglenatides, potent antiproliferative cyclic peptides isolated from the freshwater photosynthetic microalga Euglena gracilis. Angewandte Chemie International Edition. 2022;**61**(23):e202203175

[64] Ivanova JG et al. Antioxidant and anticancer potential of extracellular polysaccharide from Porphyridium aerugineum (Rhodophyta). Fermentation. 2024;**10**(5):259

[65] Magnabosco C, Santaniello G, Romano G. Microalgae: A promising source of bioactive polysaccharides for biotechnological applications. Molecules. 2025;**30**(9):2055

[66] Dimopoulou M et al. A review of the sustainability, chemical composition, bioactive compounds, antioxidant and antidiabetic activity, neuroprotective properties, and health benefits of microalgae. Biomass. 2025;**5**(1):11

[67] Nasri N et al. Enhancement of astaxanthin production in Haematococcus pluvialis using zinc oxide nanoparticles. Journal of Biotechnology. 2021;**342**:72-78

[68] Lotfi A, Abroodi Z, Khazaei M. Biological activities of astaxanthin in the treatment of neurodegenerative diseases. Neurodegenerative Disease Management. 2024;**14**(6):241-256

[69] Fassett RG, Coombes JS. Astaxanthin, oxidative stress, inflammation and cardiovascular disease. Future Cardiology. 2009;**5**(4):333-342

[70] Chen Y et al. Dietary protective potential of fucoxanthin as an active food

component on neurological disorders. Journal of Agricultural and Food Chemistry. 2023;**71**(8):3599-3619

[71] Cui S et al. Fucoxanthin alleviated atherosclerosis by regulating PI3K/AKT and TLR4/NFκB mediated pyroptosis in endothelial cells. International Immunopharmacology. 2023;**120**:110370

[72] Sayanova O, Napier JA. Metabolic engineering of microalgae for sustainable production of omega-3 long chain polyunsaturated fatty acids. Current Biotechnology. 2016;**5**(3):198-212

[73] Dyall S. Amyloid-beta peptide, oxidative stress and inflammation in Alzheimer′ s disease: Potential neuroprotective effects of Omega-3 polyunsaturated fatty acids. International Journal of Alzheimer's Disease. 2010;**2010**(1):274128

[74] Puengpan S et al. Phycocyanin attenuates skeletal muscle damage and fatigue via modulation of Nrf2 and IRS-1/AKT/mTOR pathway in exercise-induced oxidative stress in rats. PLoS One. 2024;**19**(9):e0310138

[75] Davinelli S et al. Astaxanthin as a modulator of Nrf2, NF-κB, and their crosstalk: Molecular mechanisms and possible clinical applications. Molecules. 2022;**27**(2):502

[76] Kim M-B et al. Fucoxanthin inhibits lipopolysaccharide-induced inflammation and oxidative stress by activating nuclear factor E2-related factor 2 via the phosphatidylinositol 3-kinase/AKT pathway in macrophages. European Journal of Nutrition. 2021;**60**(6):3315-3324

[77] Ávila-Román J et al. Anti-inflammatory and anticancer effects of microalgal carotenoids. Marine Drugs. 2021;**19**(10):531

[78] Matin M et al. Bioactive potential of algae and algae-derived compounds: Focus on anti-inflammatory, antimicrobial, and antioxidant effects. Molecules. 2024;**29**(19):4695

[79] Raj S et al. Microalgae as a source of mycosporine-like amino acids (MAAs); advances and future prospects. International Journal of Environmental Research and Public Health. 2021;**18**(23):12402

[80] Duan X et al. Bioaccessibility, bioavailability and bioactivities of carotenoids in microalgae: A review. Food Reviews International. 2024;**40**(1):230-259

[81] Hashimoto T et al. Pharmacokinetics of fucoxanthinol in human plasma after the oral administration of kombu extract. British Journal of Nutrition. 2012;**107**(11):1566-1569

[82] Asai A et al. Biotransformation of fucoxanthinol into amarouciaxanthin a in mice and HepG2 cells: Formation and cytotoxicity of fucoxanthin metabolites. Drug Metabolism and Disposition. 2004;**32**(2):205-211

[83] Takatani N et al. Identification and tissue distribution of fucoxanthinol and amarouciaxanthin a fatty acid esters in fucoxanthin-fed mice. Food Chemistry. 2023;**410**:135318

[84] Zhao D et al. Effects of temperature, light, and pH on the stability of fucoxanthin in an oil-in-water emulsion. Food Chemistry. 2019;**291**:87-93

[85] Fernandes V, Mamatha BS. Fucoxanthin, a functional food ingredient: Challenges in bioavailability. Current Nutrition Reports. 2023;**12**(4):567-580

[86] Ding L et al. Innovative approaches to Fucoxanthin delivery:

Characterization and bioavailability of solid lipid nanoparticles with eco-friendly ingredients and enteric coating. International Journal of Molecular Sciences. 2024;**25**(23):12825

[87] ElFar OA et al. Advances in delivery methods of Arthrospira platensis (spirulina) for enhanced therapeutic outcomes. Bioengineered. 2022;**13**(6):14681-14718

[88] Elisabeth B, Rayen F, Behnam T. Microalgae culture quality indicators: A review. Critical Reviews in Biotechnology. 2021;**41**(4):457-473

[89] Aly HF, Rizk MZ, Shams SGE. Microalgae in food and feed: Safety and toxicological aspects. In: Handbook of Food and Feed from Microalgae. Academic Press; 2023. pp. 549-565

[90] Gong Y et al. Microalgae as platforms for production of recombinant proteins and valuable compounds: Progress and prospects. Journal of Industrial Microbiology and Biotechnology. 2011;**38**(12):1879-1890

[91] Chen G et al. Molecular mechanisms of the coordination between astaxanthin and fatty acid biosynthesis in Haematococcus pluvialis (Chlorophyceae). The Plant Journal. 2015;**81**(1):95-107

[92] Hoys C et al. Unveiling the underlying molecular basis of astaxanthin accumulation in Haematococcus through integrative metabolomic-transcriptomic analysis. Bioresource Technology. 2021;**332**:125150

[93] Ng IS et al. Recent developments on genetic engineering of microalgae for biofuels and bio-based chemicals. Biotechnology Journal. 2017;**12**(10):1600644

[94] Shah F, Dixit D, Elsayed M, Behera DP. Omics approaches for algal applications. In: Value-Added Products from Algae: Phycochemical Production and Applications. Cham: Springer International Publishing; 2023. pp. 357-401

[95] Saini DK et al. Microalgal metabolomes and recent biotechnological advances for their industrial application. Microbiology Research. 2024;**15**(4):2056-2069

[96] Bas TG. Bioactivity and bioavailability of carotenoids applied in human health: Technological advances and innovation. International Journal of Molecular Sciences. 2024;**25**(14):7603

[97] Kumar N, Shukla P. Microalgal multiomics-based approaches in bioremediation of hazardous contaminants. Environmental Research. 2024;**247**:118135

[98] Wichuk K, Brynjólfsson S, Fu W. Biotechnological production of value-added carotenoids from microalgae: Emerging technology and prospects. Bioengineered. 2014;**5**(3):204-208

[99] Zafar SU, Mehra A, Jutur PP. Synthetic biology-based advanced biotechnological approach in microalgal biorefinery. In: Micro-Algae: Next-Generation Feedstock for Biorefineries: Contemporary Technologies and Future Outlook. Singapore: Springer Nature; 2022. pp. 205-230

[100] Xia D et al. Recent advancements and future perspectives of microalgae-derived pharmaceuticals. Marine Drugs. 2021;**19**(12):703

Chapter 3

Phycocyanin as a Natural Antimicrobial for Biotechnological Development

Raquel Fernandes and Joana Campos

Abstract

Antimicrobial resistance has risen over the years, and by 2050, it will be one of the leading causes of death globally. This problem has intensified the exploration of natural antimicrobial agents. Phycocyanin, a photosynthetic pigment derived from cyanobacteria, red algae, and cryptomonads, has shown antimicrobial activity against potential pathogenic microorganisms. The capacity to disrupt cell membranes and induce oxidative stress, hence reducing pathogen proliferation and survival, has increased interest in phycocyanin as a biotechnological tool for medical devices, food packaging, and cosmetics. This chapter offers an overview of the phycocyanin biological profile, highlighting its antibacterial, antifungal, and antiviral activities and mechanisms. Furthermore, it explores the therapeutic and biotechnological applications as a natural food preservative in antimicrobial packaging films, as well as in formulation strategies to increase its bioavailability and efficacy for wound care and cosmetic biomaterials. This review underscores phycocyanin's potential for the generation of innovative natural-based interventions against resistant microorganisms, including pathogenic ones.

Keywords: phycocyanin, microalgae, antimicrobial activity, therapeutic applications, biotechnology

1. Introduction

The escalating crisis of antimicrobial resistance (AMR) represents one of the most critical global public health challenges of the twenty-first century [1]. By 2050, antibiotic-resistant infections could be the leading cause of death, according to projections. The World Health Organisation (WHO) estimates that they could cause about 10 million deaths each year by that time [2, 3]. In response to this threat, the scientific and medical communities must explore innovative, sustainable therapeutic avenues beyond conventional antibiotics.

This alarming scenario is exacerbated by the widespread dissemination of multidrug-resistant (MDR) pathogens. The ESKAPE group of bacteria, which includes

Enterococcus faecium, *Staphylococcus aureus*, *Klebsiella pneumoniae*, *Acinetobacter baumannii*, *Pseudomonas aeruginosa*, and *Enterobacter* species, presents a notably higher risk [4, 5]. Closely associated with hospital-acquired infections, these pathogens acquire diverse mechanisms of multidrug-resistance and virulence, which limit treatment options. Moreover, they frequently contribute to increased mortality, morbidity, and healthcare costs worldwide [4, 5].

In addition to bacteria, AMR is also rising among fungi, viruses, and parasites, creating a complex challenge for global health. In fact, the limited classes of antifungal drugs exacerbate the rise of resistant strains in fungi, such as *Candida auris* and *Aspergillus fumigatus* [6]. Similarly, the rapid evolution of viruses, such as influenza and human immunodeficiency virus (HIV), promotes the emergence of resistant variants, challenging continuous surveillance and novel antiviral strategies [7]. In parasitic diseases, AMR in *Plasmodium* spp., responsible for malaria, and protozoa like *Giardia duodenalis* remains a major challenge to disease control in endemic regions [8, 9]. Across all these microbial groups, the relentless emergence of AMR consistently outpaces the development of new therapeutics, underscoring an urgent need for novel and effective interventions.

The inappropriate use of antimicrobials in human medicine, agriculture, and aquaculture fuels the acquisition and dissemination of AMR. The result generates a powerful selective pressure that favours resistant strains and facilitates the spread of resistance genes through horizontal gene transfer and environmental contamination [10–12]. Consequently, the AMR crisis necessitates a paradigm shift toward the discovery and development of novel therapeutic and prophylactic agents, with a renewed focus on natural alternatives that can complement or replace dwindling synthetic options [5, 10, 13].

In this context, interest in bioactive compounds from sustainable and natural sources has increased substantially. Microalgae, in particular, have emerged as a promising source of compounds with therapeutic potential [14, 15]. Among these, the light-harvesting pigment-protein phycocyanin stands out. Extracted primarily from cyanobacteria like *Spirulina*, phycocyanin is a water-soluble, non-toxic compound historically recognised for its potent antioxidant, anti-inflammatory, and neuroprotective properties, supporting its use in nutraceuticals, cosmetics, and food colouring [16, 17].

Importantly, in addition to these recognised advantages, substantial evidence indicates that phycocyanin exhibits broad-spectrum antimicrobial activity against various pathogens, including antibiotic-resistant bacteria, fungi, viruses, and parasites. This activity, along with phycocyanin's inherent biocompatibility, shows that it can be used for more than just a nutritional supplement. It also suggests that phycocyanin may be a promising lead molecule for biomedical and biotechnological advancements.

Hence, this chapter aims to provide a comprehensive and critical synthesis of phycocyanin's role as a natural antimicrobial agent. It will consolidate the current knowledge about its efficacy against clinically relevant pathogens and detail the mechanistic basis of its action, from membrane disruption to redox-mediated damage. Moreover, it explores the translational use of phycocyanin in the development of advanced biomaterials for wound healing, active packaging, and cosmetic preservation. By linking fundamental science with applied biotechnology, this review points out the potential of phycocyanin as part of a more sustainable strategy to address the growing challenge of AMR.

2. Phycocyanin characterisation and bioactive properties

Phycocyanin is a water-soluble phycobiliprotein that occurs mainly in organisms from the Cyanobacteria, Rhodophyta, and Cryptophyta phyla [18]. It is a major constituent of the light-harvesting phycobilisome complexes of *Arthrospira platensis* and *Arthrospira maxima* (commercially known as Spirulina), but it is also synthesised by genera such as *Thermosynechococcus*, *Limnothrix* spp., *Geitlerinema* spp., and the extremophilic red alga *Galdieria sulphuraria* [19]. Structurally, phycocyanin comprises a protein moiety composed of α and β polypeptide subunits that associate into αβ heterodimers and a non-protein chromophore fraction formed by phycocyanobilins, which are linear tetrapyrrolic bile-pigments covalently attached to specific cysteine residues [18]. Each αβ monomer typically carries three phycocyanobilin chromophores bound through stable thioether linkages. These monomers further assemble into $(\alpha\beta)_3$ trimers and $(\alpha\beta)_6$ hexamers, generating highly ordered oligomeric structures with defined symmetry and optical properties (**Figure 1**) [18].

The extended conjugation of the phycocyanobilin chromophores and their precise positioning within the protein scaffold increases their optical properties. This structure confers intense absorption in the orange-red region of the visible spectrum, increases fluorescence emission in the red region, and allows high molar extinction coefficients and quantum yields. These features, combined with the intrinsic water solubility and the net negative charge of the protein at physiological pH, explain the broad use of phycocyanin as a natural blue colourant and fluorescent marker. Moreover, its structure also favours phycocyanin interaction with biological membranes, macromolecules, and interfaces. While these photophysical properties are optimised for light harvesting *in vivo*, the same conjugated chromophore system also underlies the molecule's redox reactivity and its ability to participate in reactive oxygen species (ROS)-related processes outside the photosynthetic context [19]. Conformational and spectral properties of phycocyanin are, however, highly sensitive to environmental factors such as pH, ionic strength, and temperature, which can induce reversible or irreversible changes in oligomerisation state, chromophore

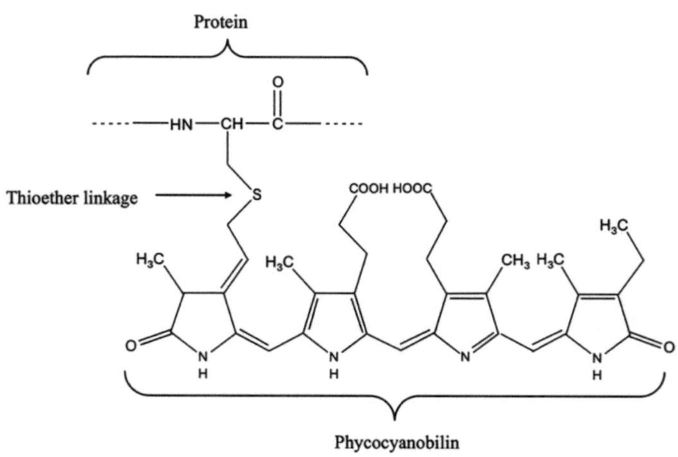

Figure 1.
Chemical structure of phycocyanin, including the protein and the non-protein chromophore (phycocyanobilin) fractions.

microenvironment and overall stability [20]. These structural perturbations are expected to negatively affect its antimicrobial performance, since both ROS generation [21] and membrane interaction [22] depend on an intact chromophore-protein architecture [18, 20].

Beyond its photosynthetic role, phycocyanin has been recognised as a valuable bioactive molecule in the food, cosmetic, and biotechnology fields due to its antioxidant, anti-inflammatory, and immunomodulatory properties [16]. As an antioxidant, phycocyanin directly scavenges reactive oxygen and nitrogen species and modulates endogenous defence enzymes. Its anti-inflammatory and immunomodulatory effects are equally significant, linked to the regulation of key signalling pathways and cytokine production [16]. These properties fundamentally converge to establish the mechanistic basis of phycocyanin's antimicrobial activity. The redox-active chromophore (capable of generating or scavenging reactive oxygen species), the surface charge (which enhances membrane interaction), and the metal-chelating capacity collectively facilitate the targeted disruption of microbial cells [21, 22].

Despite this growing body of evidence, more detailed studies on structure-function relationships and on phycocyanin's interactions with cellular targets will be crucial to uncover novel bio-functional properties, enhance its economic value and support the rational design of targeted antimicrobial strategies based on this pigment-protein complex. Thus, the structural and functional attributes detailed in this chapter provide a rational framework for understanding phycocyanin's reported antimicrobial effects against diverse pathogens, which range from membrane disruption to the induction of lethal oxidative stress.

3. Antimicrobial activity of phycocyanin

Among the 33 bacterial pathogens associated with global mortality, *S. aureus*, *Streptococcus pneumoniae*, *Escherichia coli*, *K. pneumoniae*, and *P. aeruginosa* were responsible for 54.90% of deaths and 30.90% of all infection-related deaths in 2019 [23]. These pathogens were also associated with a high number of years of life lost (YLLs) worldwide, emphasising the urgent need for new antibacterial agents. Phycocyanin has been increasingly studied for its antimicrobial properties. Extracts of phycocyanin from *Anabaena oryzae*, *Oscillatoria* spp., *Spirulina* spp., and *Geitlerinema* spp. frequently target *S. aureus* and *Bacillus* species among Gram-positive bacteria. Researchers have also reported inhibition of Gram-negative bacteria, specifically against *E. coli*, *Pseudomonas* spp., and *Klebsiella* spp. (**Table 1**).

For instance, Osman and collaborators [25] described that phycocyanin from *Anabaena oryzae*, at concentrations ranging from 10 to 100 μg/mL, effectively targets both Gram-positive (*Bacillus cereus* and *S. aureus*) and Gram-negative bacteria (*E. coli* and *K. pneumoniae*) by disrupting their cell walls and membranes. Similarly, phycocyanin from *Oscillatoria* spp. has shown inhibitory activity against *Bacillus subtilis, Streptococcus pyogenes*, *S. aureus*, *Pseudomonas* spp., and *Klebsiella* spp. [26, 27].

Numerous studies on phycocyanin extracted from *S. platensis* and *Spirulina* spp. have demonstrated efficacy against *S. aureus*, *Klebsiella* spp., *Salmonella*, and *E. coli* [28, 33, 34, 37]. In particular, the combination of phycocyanin and plasma treatment has been reported to preserve rainbow trout fillets (*Oncorhynchus mykiss*) for up to 18 days by inhibiting *S. aureus*, *Salmonella*, and *E. coli* [28].

About 10% of a phycocyanin extract from *S. platensis* inhibits *Propionibacterium acnes* and *Staphylococcus epidermidis* [29]. About two Gram-positive human skin

Microalgae species	Antibacterial		Anti-parasitic	Anti-fungal	Anti-viral	PC doses	Ref.
	Gram + inhibition	Gram − inhibition					
Anabaena	*Bacillus mucilaginosus, Sarcina lutea*	*Vibrio harveyi*, *Vibrio parahaemolyticus*	—	—	—	Crude extract	[24]
Anabaena oryzae	*Bacillus cereus, Staphylococcus aureus*	*Escherichia coli*, *Klebsiella pneumoniae*	—	—	—	10–100 μg/mL	[25]
Oscillatoria minima	*Bacillus subtilis, Streptococcus pyogenes*	*Pseudomonas fragi, E. coli, Proteus vulgaris, Klebsiella oxytoca*	—	—	—	16 μg/mL	[26]
Oscillatoria spp.	*S. aureus*	*Pseudomonas* spp., *Klebsiella* spp.	—	—	—	4–18 μg per well	[27]
Spirulina platensis	*S. aureus*	*Salmonella*, *E. coli*	—	—	—	PC and plasma 70–80 kV, 2–5 min	[28]
	Propionibacterium acnes, Staphylococcus epidermidis	—	—	—	—	10% crude extract	[29]
	Streptococcus thermophilus, *Lactobacillus bulgaricus*	—	—	—	—	Yoghurt with 2–8%	[30]
	—	—	—	*Candida albicans, Aspergillus niger, Aspergillus flavus, Penicillium* spp., *Rhizopus* spp.	—	40–80 μg/mL	[31]
	—	—	—	—	HCoV-229E	Crude extract	[32]

Microalgae species	**Antibacterial**		**Anti-parasitic**	**Anti-fungal**	**Anti-viral**	**PC doses**	**Ref.**
	Gram + inhibition	**Gram − inhibition**					
Spirulina spp.	*S. aureus*	*E. coli*, *Pseudomonas aeruginosa*, *K. pneumoniae*	—	—	—	100 µg	[33]
	—	*Salmonella* Typhimurium	—	—	—	0.5 g/L CAL- PC	[34]
Geitlerinema spp.	*Micrococcus luteus*, *S. aureus*	*E. coli*, *S.* Typhimurium	—	*Candida krusei*, *Candida glabrata*	—	250 and 500 µg	[35]
Nostoc muscorum	—	—	*Plasmodium falciparum*	—	—	~3 µg/mL	[36]
Nostoc spp. and *Spirulina* spp.	*S. aureus*	*E. coli*	—	—	—	3–25 µg/mL	[37]
Synechococcus spp.	—	*E. coli*	—	—	—	Crude extract	[38]
Phyllymenia gibesii	—	*E. coli*	—	*C. albicans*	—	0.1 mg/mL	[39]
Westiellopsis spp.	*B. subtilis*	*Pseudomonas* spp.	—	—	—	1 µg/disc	[40]

CAL-PC: chitosan-alginate-coated phycocyanin, PC: phycocyanin.

Table 1.
Antimicrobial activity of phycocyanin from different microalgae sources.

commensals that, under dysbiotic conditions, may trigger inflammation and exacerbate acne [41]. Consequently, phycocyanin from these species is being explored as a promising anti-acne therapeutic agent.

Furthermore, phycocyanin has been shown to inhibit *Streptococcus thermophilus* and *Lactobacillus bulgaricus* in yoghurt during 21-day storage. The treatment increased yoghurt pH in a dose-dependent manner, supporting its potential use as a bioactive colourant in food applications [30].

Murugan and collaborators [31] demonstrated the antifungal activity of phycocyanin from *S. platensis* against *Candida albicans*, *Aspergillus niger*, *Aspergillus flavus*, *Penicillium* spp., and *Rhizopus* spp. Moreover, phycocyanin showed antiviral activity against the human coronavirus HCoV-229E, with a selectivity index of approximately 27.9, indicating strong antiviral efficacy and low cytotoxicity [32]. In this study, phycocyanin was encapsulated into nanophytosomes and incorporated into cow's milk fermented with *Lactobacillus rhamnosus* KU985435, resulting in a stable and functional yoghurt.

Alibabaei and collaborators [33] evaluated the antibacterial effect of chitosan-alginate-coated phycocyanin extract on *Penaeus semisulcatus* shrimp infected with *Salmonella typhimurium*, showing significant inhibition. Similarly, Hajiyeva [35] reported antimicrobial activity of phycocyanin from *Geitlerinema* spp. against *S. typhimurium*, a foodborne pathogen associated with gastroenteritis in humans. Whole genome sequencing of *S. typhimurium* has revealed high genomic diversity and AMR among pathovariants, reinforcing the importance of alternative therapeutics [42]. Collectively, these studies highlight the potential of phycocyanin as an effective active molecule for gastrointestinal infections.

Phycocyanin from *Geitlerinema* spp. has demonstrated antifungal effects against *Candida krusei* and *Candida glabrata* [35]. Phycocyanin from *Nostoc muscorum* exhibited antiparasitic activity by inhibiting *Plasmodium falciparum*, the protozoan parasite responsible for the most pathogenic form of human malaria [36]. In this study, the authors suggest that phycocyanin binds to ferriprotoporphyrin-IX, inhibiting haemozoin polymerisation. Additionally, crude extracts and 0.1 mg/mL of phycocyanin from *Synechococcus* spp. and *Phyllymenia gibesii* were efficient against *E. coli* and *C. albicans* [38, 39]. Other cyanobacteria, such as *Westiellopsis* spp. and *Anabaena* spp., demonstrated broad-spectrum antibacterial activity, including inhibition of *Bacillus subtilis*, *Sarcina lutea*, *Pseudomonas* spp., *Bacillus mucilaginosus*, *Vibrio harveyi*, and *Vibrio parahaemolyticus* [24, 40].

4. Antimicrobial mechanisms of phycocyanin

Phycocyanin is widely recognised for its multifunctional biological activities, particularly its ability to modulate oxidative stress, immune responses, and microbial viability. Its antimicrobial effects arise from a complex interplay between redox regulation, metal ion interactions, membrane perturbation, and interference with bacterial communication systems [43].

Importantly, phycocyanin displays a dual redox behaviour, acting either as an antioxidant or as a pro-oxidant depending on environmental conditions, molecular partners, and physicochemical context. It was firstly recognised by its antioxidative capacity, acting as a scavenger of alkoxy, hydroxyl, and peroxyl radicals, while suppressing nitrite production and inducible nitric oxide synthase (iNOS) expression and inhibiting microsomal lipid peroxidation [16, 44]. These properties contribute to cellular protection against oxidative damage, as demonstrated in mammalian systems, where phycocyanin preserves mitochondrial membrane potential, prevents cytochrome c release, reduces ROS accumulation, and attenuates apoptosis and necrosis signalling cascades under oxidative stress conditions [45, 46]. In this context, a study showed that phycocyanin reduced mitochondrial membrane potential loss and inhibited the RIP-1/RIP-3/p-MLKL signalling cascade that triggers necrosis on GC-1 spg spermatocyte cells, protecting against H_2O_2-induced cell damage [45]. In blastocysts, phycocyanin markedly reduced apoptosis and DNA damage by preserving mitochondrial membrane potential and limiting hydrogen peroxide (H_2O_2)-induced ROS accumulation [46].

Despite this antioxidant role, phycocyanin can also exert pro-oxidant antimicrobial effects through coordination with metal ions. Phycocyanin readily binds Fe^{3+}/Fe^{2+}, forming iron-phycocyanin complexes that, under mildly acidic conditions, enhance fenton-like reactions, leading to increased production of highly reactive species [21]. This ROS amplification has been proposed as a promising antimicrobial strategy, particularly against pathogens susceptible to oxidative damage.

The interaction of phycocyanin with other metal ions further modulates its antimicrobial activity. Binding to Cu^{2+} or Ag^{+} induces conformational changes and fluorescence quenching, suggesting alterations in redox chemistry and electron transfer potential [47]. Specifically, functionalisation of silver nanoparticles (AgNPs) with 0.0197 g of phycocyanin improved the antibacterial property of AgNPs, inhibiting *P. aeruginosa* and *S. aureus* biofilm production [48]. These findings indicate that phycocyanin can act as a bioactive surface modifier, improving nanoparticle stability and antimicrobial performance.

Beyond redox-based mechanisms, phycocyanin directly interacts with bacterial cell envelopes. Its surface charge promotes electrostatic attraction to negatively charged components of bacterial membranes, which leads to membrane destabilisation. Following binding, phycocyanin can induce pore and channel formation, resulting in membrane permeability loss and bacterial cell death [22]. Supporting this mechanism, morphological analyses revealed that phycocyanin induces structural damage to bacterial cell walls rather than inhibiting protein biosynthesis, highlighting a membrane-targeted mode of action [25].

Phycocyanin also exhibits antimicrobial activity through synergistic protein–protein interactions. Lysozyme, a natural antimicrobial enzyme that hydrolyses β-(1–4)-glycosidic bonds in bacterial peptidoglycan [49, 50], has been shown to interact strongly with phycocyanin. Molecular docking and simulation studies demonstrated that this interaction is driven by electrostatic forces, hydrophobic interactions, and hydrogen bonding, involving specific α- and β-subunits of phycocyanin and key lysozyme residues. Importantly, this complex formation enhances antioxidant activity and radical inhibition, potentially amplifying antimicrobial efficacy through combined membrane degradation and oxidative stress modulation [51].

In addition, phycocyanin has emerged as an effective photosensitiser in antimicrobial photodynamic therapy. Upon light absorption, phycocyanin transitions to an excited triplet state and transfers energy or electrons to molecular oxygen, generating ROS such as H_2O_2, superoxide anion (O_2^-), and hydroxyl radicals (•OH). These reactive species disrupt microbial metabolism and damage membranes, proteins, and nucleic acids, ultimately leading to microbial death [52].

Recent evidence further expands the antimicrobial spectrum of phycocyanin by demonstrating its ability to inhibit quorum sensing in *P. aeruginosa*. Phycocyanin significantly attenuates biofilm formation, motility, pyocyanin production, and *Pseudomonas* quinolone signal synthesis, while downregulating virulence-associated genes, indicating a non-lethal but highly effective antivirulence strategy (**Figure 2**) [43].

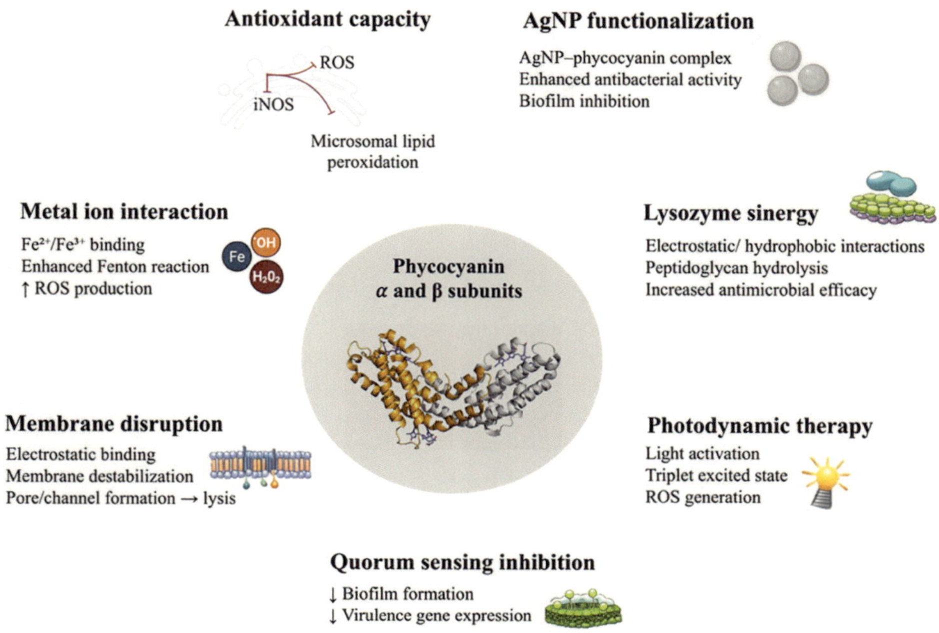

Figure 2.
Proposed mechanisms for the antimicrobial activity of phycocyanin. The α subunit of phycocyanin is shown in grey and the β subunit in yellow. AgNP: silver nanoparticles, iNOS: inducible nitric oxide synthase, ROS: reactive oxygen species.

5. Biotechnological applications of phycocyanin's antimicrobial properties

Bacterial infections present a major challenge to the healing process of wounds, particularly in chronic wounds, which are highly susceptible to contamination by pathogenic microorganisms. Due to its antimicrobial properties, phycocyanin has been explored in a wide range of biotechnological applications, including wound healing. Studies using both *in vitro* and *in vivo* models have demonstrated its capacity to enhance healing processes.

Phycocyanin from *Spirulina fusiformis* [53] and *S. platensis* [54] in concentrations ranging from 10 to 200 μg/mL promoted wound healing by upregulating urokinase-type plasminogen activator (uPA). This stimulates human fibroblast proliferation through phosphoinositide 3-kinase (PI3K) pathway induction via cyclin-dependent kinases CDK1 and CDK2, enhancing human keratinocyte proliferation and migration. These effects were similarly observed in murine and rat models using 1.25 and 2.50% phycocyanin extract and 75 μg/mL phycocyanin, significantly improving wound healing outcomes (**Table 2**) [53, 54].

The antimicrobial activity of phycocyanin has also driven the development of functional materials across a wide range of sectors, including food packaging and cosmetics. Phycocyanin has been added to eco-friendly and edible packaging films to help keep food safe and fresh.

Indeed, a biofilm composed of phycocyanin, carboxymethyl cellulose, and sodium alginate improved physical and chemical interaction with the cassava starch matrix [55]. Chentir and collaborators [56] described the development of a gelatin-based film in which gelatin established noncovalent cross-links with phycocyanin, producing a more compact film network. In another study, coffee by-products were

Objective	Product	Main composition	Ref.
Food industry	Biopolymer preservation film	PC, CMC, SA	[22]
		PC, gelatin	[55]
		PC, curcumin, lycopene, chitosan, and coffee silverskin	[56]
		Anthocyanin, PC, gelatin, soybean	[57]
		Zein, C-PC	[58]
		Chitosan and pectin from CM incorporated with procyanidins and PC	[59]
	Food emulsion	1.5% (wt) PC and 0.1% (wt) diutan gum, pH 2.5	[60]
		5% (wt) PC and 7.5% (wt) pectin	[61]
		1:1 PC/tween 80 and k-carrageenan	[62]
	Probiotic	8 mg/mL PC	[63]

Objective	Product	Main composition	Ref.
Wound dressing	Hydrogel	5% (wt) PVA, 0.5% (wt) κ-carrageenan, 0.5% (wt) PC, 0.5% (wt) borax, 0.05% (wt) rhein 2.5 mg/mL	[64]
		20 mg PC, PCNP@ND-336 and 25:75 ratio collagen/chitosan	[65]
		6% quaternised chitosan, 2% PC, 1% tetrakis HMPS, and 1% silk fibroin	[66]
		Gelatine 50% (w/v), C-PC 20% (w/v), silk fibroin 20% (w/v), tetracaine 5% (w/v), and 0.5% (w/v) rhein solution	[67]
		0.1 g PVA, 0.15 g SA, 0.5 g acrylamide, 0.05 g MBA, 0.1 g APS loaded with 1:15 (w/v) PC	[68]
		Encapsulation of 150–600 mg/100 mL PC and *Pediococcus pentosaceus* MZ413646 with whey protein nanofibrils in gelling agents	[69]
		Collagen-based hydrogels chemically crosslinked with ~0.3% (w/w) transglutaminase and 0–6 mg/mL PC	[70]
		Gelatine-based hydrogels enhanced with 1 mg/mL C-PC and nopal mucilage	[71]
		PC-templated copper sulphide nanoparticles encapsulated in alginate	[72]
		1% (w/w) PC with chitosan (3% w/w), SA (3.5% w/w), Soligel™ (1.5% w/w), HEC (5.5% w/w), NaCMC (5% w/w), HPMC (15% w/w), HEMC (3% w/w), Carbopol® Ultrez 21 (0.8% w/w), Pemulen™ (0.5% w/w), and poloxamer 407 (15% w/w)	[73]
		PVA polymer, plasticised with glycerol (~ 15% w/w), embedded with 2% (w/w) PC microparticles encapsulated by nanochitosan and SA	[74]

Objective	Product	Main composition	Ref.
Pharmaceutical	Antimicrobial photodynamic therapy	PC and 635 nm diode laser with a power of 500 mW	[75]
		PC activated with 635 nm diode laser, 220 mW, 0.34 W/cm^2	[76]
		2.4×10^{-3} mol/L chlorophyllin-PC mixture with 104 J/cm^2 of a 635 nm diode laser during 3 min	[77]
		5 mg/mL chlorophyllin-PC and 220 mW diode laser, 635 nm, 103 J/cm^2, 3 min	[78]
		2 mg/mL PC and 635 nm diode laser 34.37–171.87 J/cm^2	[79]
		125 µg/mL PC and diode laser (635 nm, 220 mW)	[80]
		20 mg/mL PC, 100 mM Na_2SeO_3, 0.5 mL of 0.4 M NaOH, 100 mM selenite/GSH	[81]
	Antiparasitic medication	PC nanoparticles with zirconia (290.2 nm, zeta potential 48.3 mV and 0.390 PDI)	[82]
Cosmetic	Skin cream	1.13% *Spirulina platensis* extract	[83]
		1.25 and 2.50 mg/mL PC	[84]
		20–25 mg/mL PC incorporated into a standard semi-solid emulsion base (vaseline/paraffin/cetostearyl/SDS), with preservatives, applied topically twice daily	[85]

APS: ammonium persulphate, CM: carboxymethyl, CMC: carboxymethylcellulose, GSH: glutathione, HEC: hydroxyethyl cellulose, HEMC: hydroxyethyl methylcellulose, HMPS: hydroxymethyl phosphonium sulphate, HPMC: hydroxypropyl methylcellulose, MBA: N, N′-methylene-bis-acrylamide, Na_2SeO_3: sodium selenite, NaCMC: sodium carboxymethyl cellulose, NaOH: sodium hydroxide, PC: phycocyanin, PDI: polydispersity index, PVA: polyvinyl alcohol, SA: sodium alginate, and SDS: sodium dodecyl sulphate.

Table 2.
Antimicrobial biotechnological applications of phycocyanin.

recycled to produce biocomposite plastic wraps, where phycocyanin increased physical interactions with chitosan [57]. Tavakoli and collaborators [58] generated an intelligent tracking film to monitor fish spoilage using phycocyanin to form strong hydrogen bonds. Additionally, interactions between c-phycocyanin and zein under an electric field and acidic conditions created an edible film with higher firmness [59]. The combination of phycocyanin and procyanidins with chitosan created higher tensile strength films with antioxidant, antibacterial, and ultraviolet barrier properties successfully used in food packaging [22].

In emulsified systems, Tello [60] described a non-aggregated and stable emulgel formulated with avocado oil with 1.50% (wt) phycocyanin and 0.10% (wt) diutan gum. Another study demonstrated that 5.00% (wt) phycocyanin and 7.50% (wt) pectin improved the physical stability of avocado and lemon oil emulgels [61]. The combination of phycocyanin and k-carrageenan has also shown promise in stabilising nanoemulgels [62].

Phycocyanin, widely known as a nutraceutical, exhibits antioxidant, anti-inflammatory, immunomodulatory, and cholesterol-lowering properties [86]. However, its direct effects on probiotic viability and growth have been less explored. Chakroun and collaborators [63] provided new evidence that phycocyanin can modulate the growth of *Limosilactobacillus reuteri*, promoting its growth and extending survival in the stationary phase. By supporting colonisation of *L. reuteri* and inhibiting pathogenic microorganisms, phycocyanin could be a key natural molecule to regulate and support a healthy gut microbiome.

Several hydrogel-based materials incorporating phycocyanin have shown promising applications in wound healing. A phycocyanin/κ-carrageenan-based hydrogel demonstrated positive haemostatic activity and biocompatibility, increasing collagen deposition and inhibiting the growth of *S. aureus* and *P. aeruginosa* [64]. Similarly, a collagen/chitosan/genipin hydrogel loaded with phycocyanin nanoparticles enhanced healing in delayed diabetic wounds [65]. Zhang and collaborators [66] synthesised antibiotic-free hydrogels containing phycocyanin and quaternised chitosan, which effectively inhibited *E. coli* and *S. aureus* and promoted wound closure. A multifunctional hydrogel containing 2% phycocyanin exhibited strong stretchability, compressibility, adhesion, and antibacterial properties against *P. aeruginosa* and *S. aureus* [67]. Phycocyanin-loaded alginate-based hydrogels also improved stiffness, resisted deformation, and allowed controlled phycocyanin release over 9 hours, highlighting their potential for drug delivery [68]. Negm El-Dein [69] reported that phycocyanin-loaded probiotic hydrogels enhanced re-epithelialisation, reduced inflammation, and supported granulation tissue formation in Wistar rats. Additionally, copper sulphide nanoparticles functionalised with commercial phycocyanin in a photo-crosslinkable methacrylated pullulan matrix promoted wound healing in mice and increased antibacterial activity against *E. coli* and *S. aureus* [72]. Recent studies have also integrated phycocyanin into biocompatible matrices that promote skin regeneration through antioxidant and antimicrobial actions [70–72]. These biomaterials exhibit enhanced mechanical and swelling behaviour, supporting cell viability and tissue repair. Galinytė and collaborators [73] developed a hydrogel composed of sodium alginate and hydroxyethyl cellulose that effectively maintained phycocyanin stability, enabled controlled release, and facilitated skin penetration. Furthermore, packaging films containing 2% phycocyanin significantly inhibited *S. aureus* growth and preserved fillet quality for up to 15 days [74].

Beyond hydrogels, phycocyanin has been widely investigated for use in antimicrobial photodynamic therapy, particularly for oral infections. When activated with a 635 nm diode laser, phycocyanin exhibited significant bactericidal effects against *Streptococcus mutans* [75, 76]. A commercial chlorophyllin-phycocyanin blend, when combined with the same laser, reduced *S. mutans gtfB* gene and GtfB protein expression [77] and reduced *S. mutans* biofilm *ex vivo* by 36.90% [78]. In endodontic models, phycocyanin alone and combined with antimicrobial photodynamic therapy reduced *Enterococcus faecalis* viability by 38.10 and 89.50%, respectively [79].

This approach also demonstrated a potential against periodontitis-associated pathogens. The combination of phycocyanin with a 635 nm diode laser significantly reduced *Aggregatibacter actinomycetemcomitans* biofilm on titanium surfaces by 40.07% [80]. Additionally, selenium nanoparticles functionalised with phycocyanin exhibited strong antimicrobial photodynamic therapy-mediated antibacterial activity against *S. aureus, E. coli*, and methicillin-resistant *S. aureus* (MRSA) [81].

In parasitic infection models, treatment of immunocompromised mice infected with *Cryptosporidium* using zirconia-based phycocyanin nanoparticles restored intestinal tissue architecture and reduced inflammation, indicating an antiparasitic effect [82].

Microalgae and their bioactive compounds, including phycocyanin, are also increasingly recognised as promising ingredients in natural cosmetic formulations [87]. Gunes [83] demonstrated that 1.13% crude *Spirulina platensis* extract, rich in phycocyanin, stimulated keratinocyte proliferation and collagen type I production, promoting wound repair. These findings are supported by other studies demonstrating that phycocyanin extract enhances cellular proliferation and tissue regeneration [54]. Dahmen and collaborators [84] reported that topical creams containing 1.25–2.50 mg/mL phycocyanin inhibited the growth of *C. albicans*, *K. pneumoniae*, and *E. faecalis*, while increasing keratinocyte cell viability. A phycocyanin-based cream also achieved 55–95% wound closure, restored epidermal structure, and reduced fungal colonisation in a *C. albicans*-infected skin wound murine model, significantly accelerating the healing process [85].

Overall, the broad-spectrum efficacy of phycocyanin against Gram-positive and Gram-negative bacteria, as well as fungi, viruses, and parasites, highlights its wound-healing properties and potential as a natural therapeutic target. Additionally, its ability to stimulate tissue regeneration, modulate probiotic growth, and provide controlled release in advanced delivery systems generates significant interest for clinical and commercial applications.

6. Patents and clinical trials using phycocyanin

Recent patents highlight the growing scientific and commercial interest in phycocyanin as an antimicrobial and protective bioactive compound, spanning applications from infectious disease prevention to oral health and neuroprotection. These intellectual property filings highlight diverse formulation strategies, routes of administration, and therapeutic targets, reflecting the versatility of phycocyanin as a functional molecule.

In the field of antiviral prophylaxis, US-2004209811-A1 discloses the use of C-phycocyanin as an oral prophylactic agent to prevent viral infections. The patent is based on the premise that phycocyanin's immunomodulatory and antioxidant properties enhance host defence mechanisms, thereby reducing their susceptibility to viral entry or replication. Similarly, patent CN-1861189-A outlines the development of a pharmaceutical extract containing phycocyanin to prevent infection and block *influenza* viral replication. This patent emphasises the inhibitory effects of phycocyanin on viral proliferation, suggesting a dual mechanism involving both direct antiviral activity and modulation of host inflammatory responses. While these patents support the prophylactic potential of phycocyanin, they rely primarily on preclinical or formulation-based claims, suggesting that there must be controlled clinical validation to substantiate antiviral efficacy in humans.

In oral healthcare, patents TW-201642884-A and TW-I587866-B propose phycobiliprotein-rich formulations containing phycocyanin for the prevention and management of periodontal disease. These inventions specifically target *Porphyromonas gingivalis* and *Aggregatibacter actinomycetemcomitans*, two main pathogens involved in periodontitis. The proposed approach combines antimicrobial activity with anti-inflammatory effects, aiming not only to suppress bacterial growth but also to mitigate gingival inflammation and prevent alveolar bone loss. These patents are particularly relevant, as they address biofilm-associated infections, a major clinical challenge in dentistry, although the exact contribution of phycocyanin relative to other formulation components remains insufficiently investigated.

Further applications are found in the field of neurology. Patents WO-2022063761-A1 and EP-4216980-A1 propose the use of phycocyanin for the prevention and treatment of peripheral neuropathies. These patents leverage phycocyanin's antioxidant and anti-inflammatory properties to protect neural tissues from oxidative damage and neuroinflammation. Although not primarily antimicrobial in focus, these filings highlight the broader cytoprotective mechanisms of phycocyanin that may indirectly support host resilience against infection-associated neural damage. Complementing these claims, US-20240024282-A1 specifies a therapeutic dosage range of 0.10–4.50 mg phycocyanin per kg body weight per day for human administration, providing valuable translational guidance for future clinical development and formulation standardisation.

Moreover, two clinical trials were developed to evaluate the therapeutic impact of phycocyanin-based supplements. A completed study (NCT07013890) evaluated a 30-day oral supplementation combining phycocyanin, palmitoylethanolamide, and selenium in patients with chronic prostatitis. The study reported improvements in the chronic prostatitis pain domain score and a reduction in elevated prostate-specific antigen (PSA) levels compared with controls. While these outcomes suggest anti-inflammatory and possibly antimicrobial benefits, the multi-component nature of the formulation complicates attribution of the observed effects specifically to phycocyanin. An ongoing clinical trial (NCT06997406, Taiwan) is currently assessing the efficacy of a C-phycocyanin-rich dietary supplement in patients with sudden sensorineural hearing loss, a condition associated with inflammation and oxidative stress. The results of this study may further clarify the therapeutic scope of phycocyanin in inflammation-related pathologies.

Taken together, these patents and clinical trials illustrate the expanding biomedical relevance of phycocyanin as a multifunctional therapeutic and preventive agent. From an antimicrobial perspective, the intellectual property landscape highlights applications targeting viral infections, oral pathogens, and biofilm-associated diseases, often leveraging phycocyanin's combined antioxidant, anti-inflammatory, and membrane-active properties. However, a major translational challenge emerging from both patents and clinical trials is the lack of standardisation in the phycocyanin source, purity, dosage, and formulation. Dosage ranges vary widely, and only a limited number of patents provide explicit quantitative guidance for human administration. Furthermore, differences between phycocyanin and other phycobiliproteins are often insufficiently addressed, potentially confounding efficacy and reproducibility. Establishing standard production methods, purity benchmarks, pharmacokinetic profiles, and mechanism-specific outcomes will be critical for regulatory acceptance and clinical scalability. Nonetheless, the growing body of intellectual property and early clinical data provides a strong foundation for future translational research and supports the development of innovative phycocyanin-based products across pharmaceutical, nutraceutical, and functional food sectors.

7. Challenges and future perspectives in the antimicrobial application of phycocyanin

Despite its broad activity and promising biotechnological applications, several scientific, technical and regulatory challenges still limit the full exploitation of phycocyanin as an antimicrobial agent.

Phycocyanin exhibits marked sensitivity to environmental stressors, including temperature fluctuations, light exposure, pH variations, and changes in ionic strength [19, 20]. These factors can induce partial protein unfolding, dissociation of the native $(\alpha\beta)_3/(\alpha\beta)_6$ oligomeric structures, and degradation of the phycocyanobilin chromophore, leading to colour loss, aggregation, and precipitation [20, 88]. Consequently, the pigment's ability to generate ROS and to interact with microbial membranes, two key mechanisms supporting its antimicrobial activity, becomes compromised [21, 22]. From an antimicrobial standpoint, this instability may shorten the effective shelf life of phycocyanin-based formulations and reduce in-use potency, particularly in products that must remain active over prolonged storage or under repeated light exposure in photodynamic applications [20, 88]. This issue is exacerbated in complex matrices such as foods, wound dressings, and cosmetic formulations, where variable pH, water activity, and the presence of surfactants or metal ions can accelerate degradation [30, 73]. While stabilisation strategies, such as encapsulation in polysaccharide-protein matrices (e.g., chitosan-alginate) or incorporation into hydrogels, can improve structural integrity and control release profiles, they often increase formulation complexity and cost. Moreover, these approaches may inadvertently affect the bioavailability and pharmacokinetic profiles of phycocyanin at the target site, requiring careful optimisation for each proposed use [22, 55, 68]. Optimised extraction and storage conditions (e.g., low temperature, near-neutral pH, and oxygen-limiting environments) have been proposed to mitigate these effects and improve pigment half-life in food-like systems [20]. However, such conditions are not always compatible with industrial process constraints and real-world product formats, which points to the importance for robust, scalable stabilisation approaches tailored to specific antimicrobial applications [19, 20, 89].

Beyond physicochemical instability, a significant translational gap exists due to the current reliance on *in vitro* evidence. Most studies employ diverse microbial strains, crude extracts, and non-standardised assay conditions, which complicates cross-laboratory comparisons and hinders the definition of clinically relevant doses or minimum inhibitory concentrations (MICs) [20, 89]. Frequently, the contribution of co-extracted compounds is not fully dissected, making it difficult to attribute observed effects solely to purified phycocyanin and to establish clear structure-activity relationships [26]. *In vivo* data remain limited and are largely confined to wound-healing or probiotic-modulation models, with a critical lack of infection models that directly evaluate pathogen clearance, host immune responses, and potential perturbations to the commensal microbiota [53–55, 63]. Pharmacokinetic and bioaccessibility studies suggest that phycocyanin is susceptible to proteolytic degradation and structural disruption in the gastrointestinal tract, which may limit its systemic antimicrobial efficacy and favour local administration or advanced delivery systems designed to protect the molecule. This indicates that systemic indications could require parenteral or topical routes or encapsulation strategies targeting specific infected tissues. Additionally, the potential for microbial adaptation (including the development of tolerance to phycocyanin, particularly in photodynamic therapy protocols under suboptimal irradiation conditions) has not been systematically evaluated, raising concerns about the long-term robustness of such interventions [75–80].

Regarding mechanistic and standardisation concerns, while phycocyanin is recognised for its ability to disrupt microbial membranes and induce oxidative stress, the relative roles of direct physical damage *versus* reactive oxygen species (ROS)-mediated pathways, as well as the modulation of these processes by environmental factors (e.g., pH, metal ion coordination, and oxygen availability), require further investigation [21, 22, 47]. Detailed mechanistic studies, supported by molecular and biophysical studies, are essential for the rational design of formulations and for predicting efficacy in different biological contexts, including biofilms and polymicrobial communities. Concurrently, the lack of standardised protocols for extraction, purification, and antimicrobial testing creates significant barriers to data comparability and regulatory assessment. Regulatory frameworks currently classify phycocyanin primarily as a natural colourant for food use (e.g., as an exempt colour additive under 21 CFR Part 73 in the United States or within broader categories of natural colour additives in the European Union) rather than as an antimicrobial agent [90–92]. For antimicrobial and biomedical claims, regulators will require robust, standardised data on long-term toxicity, immunogenicity, and off-target effects, including potential phototoxicity in PDT applications [37, 71–75, 86]. The absence of harmonised specifications for purity, acceptable contaminant levels (e.g., endotoxins and heavy metals), and stability profiles hampers the global development and comparison of phycocyanin-based antimicrobial products.

Moreover, transitioning from laboratory-scale proof-of-concept to cost-effective industrial production represents a major problem [20, 89]. Conventional methods for the cultivation, extraction, and purification of high-purity, and bioactive phycocyanin are often energy-intensive, solvent-dependent, and associated with significant pigment loss and batch-to-batch variability during scale-up. Antimicrobial and biomedical applications may demand more rigorous purity specifications than food colouring, including tight control of endotoxins and other contaminants, which can substantially increase production costs and limit competitiveness against synthetic alternatives [19, 20, 88]. Emerging biorefinery concepts, which integrate phycocyanin recovery with the valorisation of residual biomass for biofuels, fertilisers, or other bioproducts, offer a pathway toward improved sustainability and circularity by distributing costs and reducing waste [89, 93]. However, these multi-product schemes demand precise process integration to avoid compromising pigment quality and antimicrobial functionality. In parallel, metabolic engineering of cyanobacterial hosts or heterologous production in recombinant microbial chassis (e.g., *E. coli*) holds promise for enhancing yields and controlling pigment composition [93], but it introduces additional regulatory and biosafety considerations associated with genetically modified organisms [18, 93].

To translate phycocyanin's antimicrobial potential into practice, future research must adopt an integrated and multidisciplinary approach. Key priorities include the following:

1. Standardisation and rigorous modelling: development of complemented *in vitro* and *in vivo* infection models (e.g., chronic wound, biofilm, and gastrointestinal models) with consistent pharmacodynamic parameters and synergy with existing antimicrobials [25, 53, 63, 75–77];

2. Mechanistic and 'omics' insights: employing proteomic, transcriptomic, and metabolomic tools to elucidate phycocyanin's mode of action at a systems level and to monitor potential microbial adaptive responses [21, 22, 47, 48, 83];

3. Advanced delivery and formulation: engineering smart platforms (e.g., stimuli-responsive nanoparticles and biofilm-penetrating hydrogels) that protect phycocyanin from degradation, enhance its targeted delivery, and improve the efficacy of antimicrobial PDT [64–68, 72, 81];

4. Translational and regulatory convergence: comprehensive preclinical safety and pharmacokinetic studies, fostering early dialogue among researchers, industry, and regulators to establish consensus quality standards and efficient approval pathways [5, 10, 11, 90, 93].

By systematically addressing these interconnected scientific, technical, regulatory, and economic challenges, phycocyanin may evolve from a promising natural compound into a validated, effective, and sustainable alternative against AMR.

8. Conclusions

Phycocyanin, a natural pigment-protein primarily derived from cyanobacteria such as *Spirulina*, has demonstrated broad-spectrum antimicrobial activity against bacteria, fungi, viruses, and parasites. This bioactivity, supported by mechanistic insights into membrane disruption and ROS-mediated damage, determines phycocyanin as a promising candidate for diverse applications ranging from pharmaceuticals and nutraceuticals to functional foods, advanced wound-healing materials, and active packaging systems.

Its therapeutic versatility is being enhanced through innovative formulation strategies, including encapsulation, hydrogel incorporation, and nanoparticle functionalisation, which improve its stability, bioavailability, and targeted delivery. Nevertheless, the successful translation of phycocyanin into large-scale antimicrobial commercial use requires overcoming persistent challenges: validation in complex biological and infection models, standardised production and testing protocols, and the establishment of clear regulatory frameworks.

The integration of phycocyanin into next-generation antimicrobial platforms reflects a growing shift toward sustainable, multifunctional bioactive molecules. With continued interdisciplinary collaboration focusing on translational science, phycocyanin holds significant potential to address the pressing threat of AMR, improve food safety, and contribute to the use of more sustainable biomedical alternatives.

Author details

Raquel Fernandes[1] and Joana Campos[2,3]*

1 Centre for Research and Technology of Agro-Environmental and Biological Sciences, CITAB, Inov4Agro, University of Trás-os-Montes and Alto Douro, UTAD, Vila Real, Portugal

2 School of Medicine and Biomedical Sciences, University of Porto (ICBAS-UP), Porto, Portugal

3 Interdisciplinary Centre of Marine and Environmental Research (CIIMAR), Matosinhos, Portugal

*Address all correspondence to: jacampos@icbas.up.pt

References

[1] Naghavi M, Vollset SE, Ikuta KS, Swetschinski LR, Gray AP, Wool EE, et al. Global burden of bacterial antimicrobial resistance 1990-2021: A systematic analysis with forecasts to 2050. The Lancet. 2024;**404**(10459):1199-1226

[2] Tang KWK, Millar BC, Moore JE. Antimicrobial resistance (AMR). British Journal of Biomedical Science. 2023;**80**:11387

[3] Ahmed SK, Hussein S, Qurbani K, Ibrahim RH, Fareeq A, Mahmood KA, et al. Antimicrobial resistance: Impacts, challenges, and future p pects. Journal of Medicine, Surgery, and Public Health. 2024;**2**:100081

[4] Miller WR, Arias CA. ESKAPE pathogens: Antimicrobial resistance, epidemiology, clinical impact and therapeutics. Nature Reviews Microbiology. 2024;**22**(10):598-616

[5] Jadimurthy R, Mayegowda SB, Nayak SC, Mohan CD, Rangappa KS. Escaping mechanisms of ESKAPE pathogens from antibiotics and their targeting by natural compounds. Biotechnology Reports. 2022;**34**:00728

[6] Vitiello A, Ferrara F, Boccellino M, Ponzo A, Cimmino C, Comberiati E, et al. Antifungal drug resistance: An emergent health threat. Biomedicine. 2023;**11**(4):1063

[7] Aw DZH, Zhang DX, Vignuzzi M. Strategies and efforts in circumventing the emergence of antiviral resistance against conventional antivirals. Antimicrobials and Resistance. 2025;**3**(1):54

[8] Pramanik PK, Alam MN, Chowdhury DR, Chakraborti T. Drug resistance in protozoan parasites: An incessant wrestle for survival. Journal of Global Antimicrobial Resistance. 2019;**18**:1-11

[9] Karpe AV, Beale DJ, Tran CD. Intelligent biological networks: Improving anti-microbial resistance resilience through nutritional interventions to understand protozoal gut infections. Microorganisms. 2023;**11**(7):1800

[10] Muteeb G, Rehman MT, Shahwan M, Aatif M. Origin of antibiotics and antibiotic resistance, and their impacts on drug development: A narrative review. Pharmaceuticals. 2023;**16**(11):1615

[11] Singha B, Singh V, Soni V. Alternative therapeutics to control antimicrobial resistance: A general perspective. Frontiers in Drug Discovery. 2024;**4**:1385460

[12] Yuan S, Jin G, Cui R, Wang X, Wang M, Chen Z. Transmission and control strategies of antimicrobial resistance from the environment to the clinic: A holistic review. Science of the Total Environment. 2024;**957**:177461

[13] Gajic I, Kekic D, Jankovic M, Tomic N, Skoric M, Petrovic M, et al. Nature's arsenal: Uncovering antibacterial agents against antimicrobial resistance. Antibiotics. 2025;**14**(3):253

[14] Khavari F, Saidijam M, Taheri M, Nouri F. Microalgae: Therapeutic potentials and applications. Molecular Biology Reports. 2021;**48**(5):4757-4765

[15] Zuorro A, Lavecchia R, Contreras-Ropero JE, Martínez JBG, Bara

jas-Ferreira, C., Bara jas-Solano, A.F. Natural antimicrobial agents from algae: Current advances and future directions. International Journal of Molecular Sciences. 2024;**25**(21):11826

[16] Fernandes R, Campos J, Serra M, Fidalgo J, Almeida H, Casas A, et al. Exploring the benefits of phycocyanin: From spirulina cultivation to its widespread applications. Pharmaceuticals. 2023;**16**(4):592

[17] Campos J, Fernandes R, Barros AN. Phycocyanin, the microalgae. Functional Food: Upgrading Natural and Synthetic Sources. 2024;**11**:125

[18] Dagnino-Leone J, Figueroa CP, Castañeda ML, Youlton AD, Vallejos-Almirall A, Agurto-Muñoz A, et al. Phycobiliproteins: Structural aspects, functional characteristics, and biotechnological perspectives. Computational and Structural Biotechnology Journal. 2022;**20**:1506-1527

[19] Yu Z, Zhao W, Sun H, Mou H, Liu J, Yu H, et al. Phycocyanin from microalgae: A comprehensive review covering microalgal culture, phycocyanin sources and stability. Food Research International. 2024;**186**:114362

[20] Mao M, Han G, Zhao Y, Xu X, Zhao Y. A review of phycocyanin: Production, extraction, stability and food applications. International Journal of Biological Macromolecules. 2024;**280**:135860

[21] Xing Y, Li J, Fan Y, Lu S, Gu W. Coordination of iron ions with phycocyanin for an improved Fenton activity at weakly acidic ph. Spectrochimica Acta Part A: Molecular and Biomolecular Spectroscopy. 2024;**310**:123975

[22] Tie S, Zhang Q, Zhao Y, Wu Y, Liu D, Zhao L, et al. Design and preparation of novel antioxidant and antibacterial films containing procyanidins and phycocyanin for food packaging. RSC Advances. 2024;**14**(11):7572-7581

[23] Ikuta KS, Swetschinski LR, Aguilar GR, Sharara F, Mestrovic T, Gray AP, et al. Global mortality associated with 33 bacterial pathogens in 2019: A systematic analysis for the global burden of disease study 2019. The Lancet. 2022;**400**(10369):2221-2248

[24] Fan M, Liao Z, Xin Wang R, Xu N. Isolation and antibacterial activity of *anabaena* phycocyanin. African Journal of Biotechnology. 2013;**12**(15):1716-1720

[25] Osman A, Salama A, Ghany AA, Sitohy M. Antibacterial activity and mechanism of action of phycocyanin extracted from an Egyptian strain of *Anabaena oryzae* sos13. Zagazig Journal of Agricultural Biochemistry and its Application. 2015;**42**:30-321

[26] Venugopal VC, Thakur A, Chennabasappa LK, Mishra G, Singh K, Rathee P, et al. Phycocyanin extracted from *Oscillatoria minima* shows antimicrobial, algicidal, and antiradical activities: *In silico* and *in vitro* analysis. Anti-Inflammatory & Anti-Allergy Agents in Medicinal Chemistry. 2020;**19**(3):240-253

[27] Shanmugam A, Sigamani S, Venkatachalam H, Jayaraman JD, Ramamurthy D. Antibacterial activity of extracted phycocyanin from *Oscillatoria* sp. Journal of Applied Pharmaceutical Science. 2017;**7**(3):062-067

[28] Seyedalangi M, Sari AH, Nowruzi B, Anvar SAA. The synergistic effect of dielectric barrier discharge plasma and phycocyanin on shelf life of *Oncorhynchus mykiss* rainbow fillets. Scientific Reports. 2024;**14**(1):9174

[29] Nihal B, Gupta NV, Gowda D, Manohar M. Formulation and development of topical anti acne formulation of *spirulina* extract. International Journal of Applied Pharmaceutics. 2018;**10**(6):229-233

[30] Mohammadi-Gouraji E, Soleimanian-Zad S, Ghiaci M. Phycocyanin- enriched yogurt and its antibacterial and physicochemical properties during 21 days of storage. Lwt. 2019;**102**:230-236

[31] Murugan T, Radhamadhavan R. Screening for antifungal and antiviral activity of C-phycocyanin from *Spirulina platensis*. Journal of Pharmacy Research. 2011;**4**:4161-4163

[32] Soliman TN, El-Dein AN, Abd Al-Daim S, Allayeh A, Awad H, Flefil NS. Characterization of C-phycocyanin antioxidant, anti- inflammatory, anti-tumour, and anti-HCoV-229E activities and encapsulation for implementation in an innovative functional yogurt. Heliyon. 2024;**10**(11):e09455

[33] Sarada DV, Sreenath Kumar C, Rengasamy R. Purified C-phycocyanin from *Spirulina platensis* (nordstedt) geitler: A novel and potent agent against drug resistant bacteria. World Journal of Microbiology and Biotechnology. 2011;**27**(4):779-783

[34] Alibabaei M, Metcalf JS, Nowruzi B. Influence of coated phycocyanin on shelf life of infected *Penaeus semisulcatus*. Aquaculture, Fish and Fisheries. 2024;**4**(6):70015

[35] Hajiyeva S, Cankilic MY, Kilic V, Gorgulu S, Patrignani F, Lanciotti R. An insight into the potential antioxidant, anticancer and antimicrobial activities of *Geitlerinema* sp. C-phycocyanin extracts. Algal Research. 2025;**85**:103836

[36] Pankaj PP, Seth RK, Mallick N, Biswas S. Isolation and purification of C-phycocyanin from *Nostoc muscorum* (cyanophyceae and cyanobacteria) exhibits antimalarial activity *in vitro*. Journal of Advanced Laboratory Research in Biology. 2010;**1**:86-91

[37] Ahmadi A, Anvar SAA, Nowruzi B, Golestan L. Effect of phycocyanin and phycoerythrin on antioxidant and antimicrobial activity of refrigerated low-fat yogurt and cream cheese. Scientific Reports. 2024;**14**(1):27661

[38] Lin J-Y, Tan S-I, Yi Y-C, Hsiang C-C, Chang C-H, Chen C-Y, et al. High-level production and extraction of C-phycocyanin from cyanobacteria *Synechococcus* sp. PCC7002 for antioxidation, antibacterial and lead adsorption. Environmental Research. 2022;**206**:112283

[39] El-Din NGS, Hafez MSAE, El-Wahab MGA, Ibrahim HA. Biological activities of derived pigments and polyphenols from the newly recorded alga *Phyllymenia gibbesii*. Scientific Reports. 2024;**14**(1):21284

[40] Sabarinathan K, Ganesan G. Antibacterial and toxicity evaluation of C-phycocyanin and cell extract of filamentous freshwater cyanobacterium–*Westiellopsis* sps. European Review for Medical & Pharmacological Sciences. 2008;**12**(2):107-114

[41] Fournière M, Latire T, Souak D, Feuilloley MG, Bedoux G. *Staphylococcus epidermidis* and *Cutibacterium acnes*: Two major sentinels of skin microbiota and the influence of cosmetics. Microorganisms. 2020;**8**(11):1752

[42] Perez-Sepulveda BM, Hinton JC. Microbe profile: *Salmonella* Typhimurium: The master of the art of adaptation. Microbiology. 2025;**171**(1):001521

[43] Ying J, Guo S, Pan Y, Li D, Tai Y, Li Y, et al. The role of cyano-phycocyanin as a quorum sensing inhibitor to attenuate *Pseudomonas aeruginosa* virulence. Frontiers in Cellular and Infection Microbiology. 2025;**15**:1624927

[44] Silva BR, Silva JR. Mechanisms of action of non-enzymatic antioxidants to control oxidative stress during in vitro follicle growth, oocyte maturation, and embryo development. Animal Reproduction Science. 2023;**249**:107186

[45] Dong X, Yang F, Xu X, Zhu F, Liu G, Xu F, et al. Protective effect of C-phycocyanin and apo-phycocyanin subunit on programmed necrosis of GC-1 SPG cells induced by H_2O_2. Environmental Toxicology. 2022;**37**(6):1275-1287

[46] Niu Y-J, Zhou W, Guo J, Nie Z-W, Shin K-T, Kim N-H, et al. C-phycocyanin protects against mitochondrial dysfunction and oxidative stress in parthenogenetic porcine embryos. Scientific Reports. 2017;**7**(1):16992

[47] Bellamy-Carter J, Sound JK, Leney AC. Probing heavy metal binding to phycobiliproteins. The FEBS Journal. 2022;**289**(15):4646-4656

[48] Chegini Z, Shariati A, Alikhani MY, Safaiee M, Rajaeih S, Arabestani M, et al. Antibacterial and antibiofilm activity of silver nanoparticles stabilized with C-phycocyanin against drug-resistant *Pseudomonas aeruginosa* and *Staphylococcus aureus*. Frontiers in Bio-engineering and Biotechnology. 2024;**12**:1455385

[49] Nawaz N, Wen S, Wang F, Nawaz S, Raza J, Iftikhar M, et al. Lysozyme and its application as antibacterial agent in food industry. Molecules. 2022;**27**(19):6305

[50] Naveed M, Sajjad A, Ali A, Chan MWH, Wang F. Antibacterial and antibiofilm properties of lysozyme extracted from *Catostylus perezi*, an invaluable jellyfish bloom. The Microbe. 2025;**7**:100365

[51] Bai Y, Wang Y, Li X, Jin J, Lu Y, Jiang S, et al. Interaction mechanism and binding mode of phycocyanin to lysozyme: Molecular docking and molecular dynamics simulation. Food Chemistry. 2024;**438**:138001

[52] Polat E, Kang K. Natural photosensitizers in antimicrobial photodynamic therapy. Biomedicine. 2021;**9**(6):584

[53] Madhyastha H, Radha K, Nakajima Y, Omura S, Maruyama M. UPA dependent and independent mechanisms of wound healing by C-phycocyanin. Journal of Cellular and Molecular Medicine. 2008;**12**(6b):2691-2703

[54] Gur CS, Erdogan DK, Onbasılar I, Atilla P, Cakar N, Gurhan ID. *In vitro* and *in vivo* investigations of the wound healing effect of crude spirulina extract and C-phycocyanin. Journal of Medicinal Plants Research. 2013;**7**(8):425-433

[55] Akhtar HMS, Zhao Y, Li L, Shi Q. Novel active composite films based on carboxymethyl cellulose and sodium alginate incorporated with phycocyanin: Physico-chemical, microstructural and antioxidant properties. Food Hydrocolloids. 2024;**147**:109440

[56] Chentir I, Kchaou H, Hamdi M, Jridi M, Li S, Doumandji A, et al. Biofunctional gelatin-based films incorporated with food grade phycocyanin extracted from the *Saharian cyanobacterium Arthrospira* sp. Food Hydrocolloids. 2019;**89**:715-725

[57] Liu X, Sun H, Leng X. Coffee silverskin cellulose-based composite film with natural pigments for food

packaging: Physicochemical and sensory abilities. Food. 2023;**12**(15):2839

[58] Tavakoli S, Mubango E, Tian L, ŃDri YB, Tan Y, Hong H, et al. Novel intelligent films containing anthocyanin and phycocyanin for nondestructively tracing fish spoilage. Food Chemistry. 2023;**402**:134203

[59] Hajjari MM, Golmakani M-T, Sharif N. Electrospun zein/C-phycocyanin composite: Simulation, characterization and therapeutic application. Food Hydrocolloids. 2023;**140**:108638

[60] Tello P, Santos J, Perez-Puyana VM, Romero A, Trujillo-Cayado LA. Characterization of emulgels formulated with phycocyanin and diutan gum as a novel approach for biocompatible delivery systems. International Journal of Biological Macromolecules. 2024;**268**:131599

[61] Tello P, Calero N, Santos J, Trujillo-Cayado LA. Development of avocado and lemon oil emulgels based on natural products: Phycocyanin and pectin. Pharmaceutics. 2023;**15**(8):2067

[62] Tello P, Santos J, Calero N, Trujillo-Cayado LA. Formulation and characterization of sustainable algal-derived nanoemulgels: A green approach to minimize the dependency on synthetic surfactants. Polymers. 2024;**16**(2):194

[63] Chakroun I, Haddaji N, Fedhila K, Maatallah M, Mzoughi R, Chaabouni Y, et al. *In vitro* characterization of *Limosilactobacillus reuteri* Lac IB01 (OL468126. 1) isolated from traditional sheep dry sausage and evaluation of the activity of arthrospira platensis or phycocyanin on its growth-promoting ability. Fermentation. 2023;**9**(3):248

[64] Zhang R, Wang L, Meng L, Shang W, Ren Y, Qi Q, et al. A slime-inspired phycocyanin/κ-carrageenan-based hydrogel bandage with ultra-stretchability, self-healing, antioxidative, and antibacterial activity for wound healing. International Journal of Biological Macromolecules. 2025;**289**:138786

[65] Li Z, Qian C, Zheng X, Qi X, Bi J, Wang H, et al. Collagen/chitosan/genipin hydrogel loaded with phycocyanin nanoparticles and ND-336 for diabetic wound healing. International Journal of Biological Macromolecules. 2024;**266**:131220

[66] Zhang S, Huang C, Sun Z, Jiao P, Ding B, Ding L, et al. C-phycocyanin and quaternized chitosan based antibiotic-free hydrogels with antioxidant and antibacterial activity for wound healing. International Journal of Biological Macromolecules. 2025;**302**:140647

[67] Meng Z, Liu J, Zhang R, Ren Y, Qi Q, Cui B, et al. Phycocyanin-based multifunctional hydrogel with self-healing, hemostatic, antioxidative, and antibacterial activity for wound healing. International Journal of Biological Macromolecules. 2025;**310**(Pt 2):143254

[68] Buliga D-I, Mocanu A, Rusen E, Diacon A, Toader G, Brincoveanu O, et al. Phycocyanin-loaded alginate-based hydrogel synthesis and characterization. Marine Drugs. 2024;**22**(10):434

[69] Negm, El-Dein A, Soliman TN, Ezzat A, Abd El-Fattah MA, Aly HF, Younis EA, et al. Innovative hydrogel formulation combining phycocyanin and probiotic for enhancing skin regeneration and accelerated wound healing: A preclinical investigation in wistar rats. Probiotics and Antimicrobial Proteins. 2025;**17**

[70] Qu J, Geng X, Qin S, Li J, Li W. The utilization of TGase-crosslinked phycocyanin-collagen hydrogels

for the purpose of facilitating skin wound healing. International Journal of Biological Macromolecules. 2025;**316**:145932

[71] Antonio-Marcos EG, Carrillo Escalante HJ, Hernández-Vázquez L, Castillo Gamboa GA, Cervantes Uc JM, Claudio Rizo JA, et al. Enhancing gelatin hydrogels: Synergistic effects of C-phycocyanin and nopal mucilage on physicochemical and biocompatibility properties. Journal of Biomaterials Applications. 2025;**40**(4)

[72] Dong J, Lang Y, He J, Cui J, Liu X, Yuan H, et al. Phycocyanin-based multifunctional microspheres for treatment of infected radiation-induced skin injury. Biomaterials. 2025;**317**:123061

[73] Galinyte D, Bernatoniene J, Žilius M, Rysevaite-Kyguoliene K, Savickas A, Karosiene J, et al. *In vitro* study of cyano-phycocyanin release from hydrogels and *ex vivo* study of skin penetration. Pharmaceuticals. 2024;**17**(9):1224

[74] Nowruzi B, Ghorbani M, Ghane M, Ashrafizadeh M, Barandak R, Targholizadeh M. Studying the effect bio-based sensorns packaging based on magnetic fields extraction of phycocyanin on the shelf life of fish fillets infected with *Staphylococcus aureus*. BMC Microbiology. 2025;**25**(1):263

[75] Abbasianfar Z, Seraj B, Afrasiabi S, Fard MJK, Ghadimi S, Chiniforush N. The effect of antimicrobial photodynamic therapy using different concentrations of phycocyanin against *Streptococcus mutans*. Photodiagnosis and Photodynamic Therapy. 2024;**50**:104384

[76] Alanazi AM, Khan AA, Mahmood A, Kamal MA, Baig EA. The antibacterial efficiency of chitosan photodynamically activated phycocyanin, and Morinda oleifera against *S. Mutans* and the bonding strength between composite resin and caries-affected dentin. Pakistan Journal of Medical Sciences. 2024;**40**(5):962

[77] Afrasiabi S, Pourhajibagher M, Chiniforush N, Aminian M, Varaei SSR, Bahador A. Effects of sub-lethal dose of antimicrobial photodynamic therapy on major virulence traits of *Streptococcus mutans*. Photodiagnosis and Photodynamic Therapy. 2020;**32**:102044

[78] Afrasiabi S, Pourhajibagher M, Chiniforush N, Aminian M, Bahador A. Anti-biofilm and anti-metabolic effects of antimicrobial photodynamic therapy using chlorophyllin-phycocyanin mixture against *Streptococcus mutans* in experimental biofilm caries model on enamel slabs. Photodiagnosis and Photodynamic Therapy. 2020;**29**:101620

[79] Pourhajibagher M, Chiniforush N, Bahador A. Antimicrobial action of photoactivated C-phycocyanin against *Enterococcus faecalis* biofilms: Attenuation of quorum-sensing system. Photodiagnosis and Photodynamic Therapy. 2019;**28**:286-291

[80] Etemadi A, Bayati SE, Pourhajibagher M, Chiniforush N. *In vitro* effect of antimicrobial photodynamic therapy with phycocyanin on *Aggregatibacter actinomycetemcomitans* biofilm on SLA titanium discs. Photodiagnosis and Photodynamic Therapy. 2020;**32**:102062

[81] Xu Z, Jin Y, Zhang Z, HePeng Y, Zhang C, Yang S, et al. Phycocyanin functionalized selenium nanoparticles for type I photodynamic antibacterial therapy and wound healing. Journal of Materials Chemistry B. 2025;**13**:8208-8218

[82] Moneer EA, Akl SH, Shahin YH, Shahin SH, Elwakil BH, Eskandrani A, et al. The antiparasitic effect of C-phycocyanin nanoparticles on cryptosporidiosis in immunosuppressed mice. Journal of Parasitic Diseases. 2025;**49**(1):173-185

[83] Gunes S, Tamburaci S, Dalay MC, Deliloglu Gurhan I. *In vitro* evaluation of *Spirulina platensis* extract incorporated skin cream with its wound healing and antioxidant activities. Pharmaceutical Biology. 2017;**55**(1):1824-1832

[84] Dahmen I, Chtourou H, Hadrich F, Baccar N, Sayadi S, Ayadi H, et al. *Phormidium versicolor* PC skin cream evaluation for its stability and biological activities. Applied Microbiology and Biotechnology. 2024;**108**(1):541

[85] Zamani N, Fazilati M, Salavati H, Izadi M, Koohi-Dehkordi M. The topical cream produced from phycocyanin of *Spirulina platensis* accelerates wound healing in mice infected with *Candida albicans*. Applied Biochemistry and Microbiology. 2020;**56**(5):583-589

[86] Fernández-Rojas B, Hernández-Juárez J, Pedraza-Chaverri J. Phycocyanin a nutraceutical with antioxidant activity. In: Improving Health and Nutrition through Bioactive Compounds. Amsterdam, Netherlands: Elsevier; 2025. pp. 67-78

[87] Ramos SdP, Burck M, Costa SFFd, Assis M, Braga ARC. Spirulina as a key ingredient in the evolution of eco-friendly cosmetics. Biotech. 2025;**14**(2):41

[88] Fernández-Rojas B, Hernández-Juárez J, Pedraza-Chaverri J. Nutraceutical properties of phycocyanin. Journal of Functional Foods. 2014;**11**:375-392

[89] Kuhnholz J, Glockow T, Siebecke V, Le AT, Tran L-D, Noke A. Comparison of different methods for extraction of phycocyanin from the *Cyanobacterium Arthrospira maxima* (*Spirulina*). Journal of Applied Phycology. 2024;**36**(4):1725-1735

[90] European Food Safety Authority (EFSA) panel on food additives and nutrient sources added to food. Guidance for submission for food additive evaluations. EFSA Journal. 2012;**10**(7):2760

[91] Berg AS. Food safety and synthetic chemicals-an international comparison. Indiana International & Comparative Law Review. 2025;**35**:259

[92] Chavarráa-Fernández SM, Jiménez-Alvarado R, Santos-López EM, Hernández-HernandezAA,Cariño-CortésR. Iron nanoparticles as food additives and food supplements, regulatory and legislative perspectives. Food Science and Biotechnology. 2024;**33**(6):1295-1305

[93] Zhao X, Gao H, Wang Y, Wang Z, Zhou J. Efficient synthesis of phycocyanobilin by combinatorial metabolic engineering in *Escherichia coli*. ACS Synthetic Biology. 2022;**11**(6):2089-2097

Section 3

Innovative and Integrated Biotechnological Applications of Microalgae

Chapter 4

Microalga-Based Electricity Production

Wid Alrashidi

Abstract

Microalgae are becoming increasingly identified as a novel solution for the production of clean electricity. The chapter discusses their twin function as biological solar panels and biochemical factories, being highly capable of using sunlight and waste feedstocks to produce sustainable energy. The chapter starts with a description of the biological characteristics that render microalgae particularly suitable for energy applications, including energy-rich macromolecules and light-absorbing pigments. In addition, emerging cultivation methods—from open ponds to sophisticated photobioreactors—are assessed for their potential to optimise growth while, at the same time, reducing the use of resources. The following chapter presents a comparison and contrast of two varied electricity generation approaches: direct systems exemplified by microbial fuel cells utilising bioelectrochemical processes and indirect systems that transesterify algal biomass to biodiesel, biogas, or hydrogen for energy use. These technologies have in-built environmental benefits, for example, wastewater treatment and carbon sequestration from industrial exhausts. Despite this, there are still difficulties in the sustainable economic development of the technologies, most conspicuously in balancing energy input and output and optimising large-scale agronomy. Future developments in genetic manipulation and bioreactor technology are expected to increase both efficiency and cost-effectiveness. This chapter outlines the potential of microalgae to enhance renewable sources of energy, alongside addressing critical issues of the environment. It does this by reconciling key concepts in science and practical implementations and thus presenting a motivating vision towards sustainable development for the twenty-first century.

Keywords: microalgae, sustainability, bioenergy, renewables, biophotovoltics

1. Introduction

The quest for sustainable and renewable energy sources has escalated in recent decades, motivated by apprehensions over climate change, energy security, and the exhaustion of fossil fuel reserves. Microalgae have emerged as a viable alternative, attracting considerable interest as a possible feedstock for biofuel generation and other bioenergy applications [1]. The investigation of microalgae as an energy source is not novel; research and development have been ongoing for several decades, with the first studies concentrating on the capacity of algae to generate lipids suitable for biodiesel production [2]. The historical evolution of microalgae in renewable energy applications began in the 1970s, amid the oil crisis, when the United States Department of Energy launched the Aquatic Species Program to investigate alternate fuel sources, including microalgae [3]. This programme established the foundation for later research, examining the diverse facets of microalgae farming, lipid production, and biofuel conversion technologies. Despite the initial potential, the study was ultimately terminated due to financial and technological limitations; however, the foundational concepts and insights produced during this time have been crucial in informing subsequent research initiatives.

Microalgae offer numerous benefits compared to conventional biomass sources for bioenergy generation, making them a compelling alternative to traditional crops. A notable advantage is their rapid growth rates, with many species capable of doubling their biomass within hours. This accelerated growth yields higher outputs per unit area compared to terrestrial crops, thereby minimising the land footprint required for biofuel production. Microalgae exhibit significant photosynthetic efficiency, with certain strains converting up to 9% of solar energy into biomass. This markedly surpasses the photosynthetic efficiency of most terrestrial plants, which typically ranges between 1 and 3%. Furthermore, microalgae cultivation does not compete with food crop production, as it can be grown on non-arable land and utilise non-potable water sources, such as wastewater and seawater [4]. Microalgae biomass comprises vital macromolecules, including polysaccharides, proteins, and lipids, which are essential for energy, feed, and food production, with certain strains containing over 90% of these constituents [5]. The ability to use various water sources and non-arable land mitigates concerns about water scarcity and land use conflicts associated with traditional biofuel crops.

Microalgae are presently investigated as a bioethanol source, with the carbohydrates produced typically existing as starch, which can be readily transformed into simple reducing sugars [6]. Microalgal lipids can be transformed into biodiesel through transesterification, a recognised process that entails the reaction of lipids with an alcohol in the presence of a catalyst. The present global landscape for microalgae-derived power generation remains nascent, with merely a few commercial-scale facilities now operational. There is increasing interest in the development of microalgae-based power generation technologies, including the integration of microalgae cultivation with wastewater treatment to produce biogas for electricity generation. Recent studies have indicated the application of microalgae in fuel cells, leading to the development of innovative systems utilising biological photovoltaic devices (BPVs) [7]. Microalgae serve as catalysts in microbial fuel cells to produce energy; these fuels are eco-friendly and do not emit additional pollutants, with the generated electricity utilised to power various equipment.

The advancement of microalgae biofuels is propelled by environmental and economic considerations. From an ecological standpoint, microalgae possess the capacity to mitigate greenhouse gas emissions by sequestering carbon dioxide during photosynthesis. Microalgae can be utilised for wastewater treatment by eliminating nutrients like nitrogen and phosphorus, hence mitigating the risk of eutrophication in aquatic environments. The use of algae serves as an effective means to mitigate air pollution, as algae generate around 50% of the Earth's oxygen *via* photosynthesis [8]. Microalgae bioenergy presents an opportunity to diversify energy sources, diminish reliance on fossil fuels, and provide employment in the biotechnology and renewable energy industries. In the last 10 years, awareness of sustainability has grown, leading to stricter restrictions to reduce greenhouse gas emissions as nations strive for carbon neutrality by 2050.

2. Biological foundations of microalga-based electricity

2.1 Historical development of microalgae in renewable energy applications

The incorporation of microalgae into renewable energy systems originates from the U.S. Department of Energy's Aquatic Species Program (1978–1996), which initiated algal biofuel research utilising waste CO_2 from coal plants [9]. The overall objective of the programme was to screen and develop microalgal species with high lipid content, high growth rates, and environmental stress tolerance, traits that are critical for the economically viable production of biofuels. The work encompassed strain selection, optimisation of cultivation methods, lipid extraction, and conversion processes to transform algal biomass into fuels like biodiesel and bio-oil. The Aquatic Species Program was committed to the utilisation of waste CO_2 as a carbon source, thus addressing the requirement for renewable energy, along with exploring a possible method for mitigating greenhouse gas emissions from fossil fuel power stations. The legacy of the programme reaches not only its immediate findings but also the trajectory of subsequent research, encouraging the further exploration of the potential applications of microalgae across a range of industries, including wastewater treatment, the production of nutraceuticals, and the creation of high-value products. Algae have been investigated for a variety of uses related to food and bioactive molecules, and they have specialised mechanisms for capturing noxious gases while assimilating sunlight, nitrogen, and phosphorus into biomass [10]. The Aquatic Species Program's research not only sought to produce biofuels from microalgae but also highlight their potential as a valuable biomass resource. This encouraged new developments in algal biotechnology and integrated biorefinery concepts. Microalgae hold promise as tools to mitigate air pollution and can serve as an effective alternative to both intermittent renewable energy and combustion-based fossil fuels, making them a viable option for advanced biofuel conversions [8].

Despite initial commercialisation challenges due to economic factors, a resurgence of interest emerged in the 2000s, driven by rising oil prices, which sparked innovations in photobioreactor design and metabolic engineering [9, 11]. Recent advancements, such as CRISPR-based strain optimisation and hybrid farming technologies, have positioned microalgae as third-generation biofuel feedstocks capable of fulfilling global energy needs. Genetically engineered *Phaeodactylum tricornutum* now achieves a biomass productivity of 70.83 mg/L/h, underscoring progress in scalable production [12]. Although CRISPR technology has the potential to improve microalgae's photosynthetic efficiency, various restrictions limit its effective application. The complex taxonomic diversity of microalgae complicates the development of standardised transformation methods and the identification of suitable selection markers, thus hindering the effectiveness of genome editing. Additionally, the persistent issue of low transformation efficiency, particularly in underutilised microalgae, coupled with concerns about off-target effects and the disruption caused by the microalgae's inherent silencing mechanisms, presents significant obstacles that must be addressed to effectively employ CRISPR techniques for the precise enhancement of microalgae's photosynthetic potential [13].

2.2 Fundamental advantages of conventional biomass sources

Microalgae possess significant potential for bioenergy generation owing to their superior biological and ecological attributes relative to terrestrial crops [12]. Their photosynthetic efficiency varies between 4.5 and 9%, markedly surpassing the 1–2% efficiency of the majority of terrestrial plants, such as maize (1.8–2.2%). This efficiency arises from specialised light-harvesting complexes and carbon-concentrating processes, including carboxysomes in cyanobacteria, which reduces photorespiratory losses and improves CO_2 fixation rates [14]. Biomass productivity represents a significant advantage: microalgae may produce up to 50 times more lipids per hectare than traditional crops, with species such as *Nannochloropsis* accumulating 50% lipids by dry weight. Annual energy yields from algal biomass (145 metric tonnes dry weight per hectare) exceed those of maize by a factor of 16, even when considering shorter growth seasons—strains like *Synechococcus* sp. PCC 11901 attained doubling times of approximately 2 hours and biomass densities of 33 g/L under ideal conditions, facilitating rapid expansion. Microalgae growth necessitates no arable land and flourishes in saline water or wastewater, hence obviating conflict with food production. *Dunaliella salina* thrives in hypersaline conditions, whereas *Chlorella fusca* sequesters CO_2 at a rate of 216.2 mg/L/day when grown with CO_2-adsorbing nanofibers. These systems also facilitate carbon sequestration, with a CO_2 fixation efficiency of 10%, much surpassing terrestrial plants (1–4%) [15]. Genetic modifications further improve these characteristics. CRISPR-Cas9 modification in *Phaeodactylum tricornutum* enhances lipid retention, whereas metabolic engineering refines light absorption and carbon fixation pathways. These developments establish microalgae as scalable and sustainable bioenergy platforms that can tackle global energy and environmental issues. By using these benefits, microalgae-based systems provide a dual solution: decreasing dependence on limited fossil fuels while alleviating greenhouse gas emissions *via* integrated biorefineries and wastewater treatment [15].

3. Cultivation technologies and systems

Microalgae possess a range of intrinsic advantages compared to terrestrial biomass, making them an attractive candidate for the next generation of bioenergy production. Their photosynthetic performance is particularly higher than most land plants, as microalgae can convert 8–10% of solar energy into biomass in optimal conditions, compared to the 1–2% efficiency of conventional terrestrial crops like corn. In reality, microalgae commonly reach photosynthetic efficiencies of about 4.5%, which is still a major upgrade compared to land plants. This superior efficiency is due to their special cellular structure, highly efficient light-harvesting complexes and efficient carbon fixation pathways, which facilitates rapid biomass production [12].

Concerning productivity, microalgae can fix carbon dioxide 10–50 times faster than terrestrial plants, achieving values for annual biomass productivities from 127 to 300 t/ha—well beyond those of traditional crops. Strains as *Synechococcus sp.* PCC 11901, capable of doubling its biomass in ~2 hours and for which a biomass concentration of up to 33 gL^{-1} in the effluent, was reached under optimised conditions [16]. In addition, microalgae grow in marginal land, and the potential use of salt water or nutrient-rich wastewater means it does not compete with a food crop and minimises the need for freshwater. This ecological versatility made them resistant and resource-efficient [17].

However, microalgae are also functionally diverse at the biochemical level, offering proteins, pigments, and antioxidants in addition to lipids as the source of production for biofuels, conducive to a multi-product biorefinery concept. The recent development of genetic tools has enabled gene silencing approaches to further increase the lipid accumulation in *P. tricornutum* by blocking the activity of the key enzymes of the lipid breakdown, opening possibilities for targeted strain improvements to increase the biofuel yield. Together, these qualities highlight the disruptive nature of microalgae as a bioenergy commodity due to situations of high performance, resource economy, and environmental advantages when compared to traditional biomass analogues [18].

4. Indirect electricity production *via* microalgal biofuels

Microalgal biofuels allow renewable electricity generation through diverse conversion routes, which exploit the biochemical composition of algal biomass. New developments point to significant advancements in the production of biodiesel, biogas, and biohydrogen, except for the limitation in scale and the cost of the process.

4.1 Production and utilisation of biodiesel

Biodiesel is produced by the transesterification of lipid-containing materials such as triacylglycerides extracted from microalgae and used as fuel for diesel gensets. Strains such as *Nannochloropsis* produce 20–50% lipids, although wet biomass processing can lower the energy-intensive dewatering costs [19]. Recent innovations include the one-step extraction-transesterification with heterogeneous acid catalysts, which are important for reducing the cost of biodiesel production to a price cheaper than $4/GGE (gasoline gallon equivalent) [20].

Wastewater-based cultivation decreases nutrient costs, utilising lipid productivity of 70 mg/L/day. Biodiesel-fuelled turbine output is 40–45% electrical efficiency, but the net energy ratio (1.5–2.0) is lower than that of fossil-type fuels [21].

4.2 Biogas and biomethane pathways

Anaerobic digestion (AD) of microalgal biomass has the potential to generate energy in the form of methane-rich biogas (55–75% CH_4) that is suitable for combined heat and power (CHP) applications. Key developments include [19]:

- Co-digestion with wheat straw enhanced MCS by 77% (0.21 L/g VS) due to the synergistic C/N ratio [22].
- Thermo-alkaline pretreatment for biodegradability enhancement leading to a 46% increase in CH_4 yield.
- These systems also help in carbon sequestration, through which AD plants trap 60% of CO_2 while digesting the tons of waste food.
- And, finally, AD at an industrial scale achieves 12 tons. CO_2/acre/year sequestration but suffers from residue management issues [22].

4.3 Biohydrogen and novel energy carriers

Biohydrogen is produced using the photobiological or fermentation pathway:

a. Direct photolysis produces 0.015–1.084 mmol H_2 L^{-1} h^{-1}, but oxygen-tolerant hydrogenase engineering is necessary [23].

b. Fermentation of *Chlorella sp.* biomass of 11.65 ml H_2/L on crude glycerol with lipid co-productivity >40% [24].

Thermochemical gasification yields syngas (H_2/CO) at $7.24/kg H_2, but electrochemical pathways offer higher purity. Biohydrogen also serves as the fuel for solid oxide fuel cells (60% efficiency), but costs are generally 2–3× higher than for natural hydrogen [21].

4.4 Environmental and economic co-benefits of waste valorisation

Integrating wastewater treatment and flue-gas CO_2 utilization (with a potential of 268 million tons annually) as co-processing strategies to reduce microalgal cultivation costs [20].

- *Biorefinery concepts*: Protein ($674/ton) co-product credits biofuel production, yielding <$4/gge fuels.
- *Emission reduction*: Wind/solar power reduces emissions 50–90% relative to fossil fuels.

4.5 Persistent challenges

As shown in the diagram in **Figure 1**, persistent challenges of indirect microalgae electricity production include the following areas:

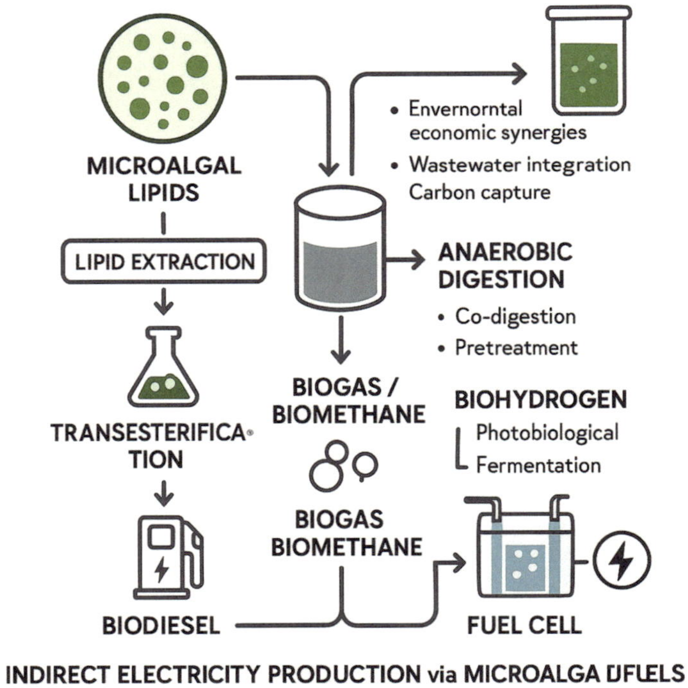

Figure 1.
Pathways for indirect electricity production from microalgal biofuels: Microalgal biomass is converted into biodiesel, biogas, and biohydrogen through various biochemical and thermochemical processes. These biofuels are then used in engines, turbines, or fuel cells to generate renewable electricity, with integrated environmental benefits such as carbon sequestration and wastewater treatment. Recent innovations and persistent challenges are highlighted across each conversion route.

- *Dewatering power*: 20–30% of overall production costs [19].
- *Inhibitors*: AD stability is decreased by ammonia and long-chain fatty acids.
- *Scale-up*: Photobioreactor fouling & raceway pond contamination hinder commercialisation [20].
- *Genetic engineering* (like CRISPR-aided lipid yields) and AI-assisted cultivation parameter optimisation are important frontiers for enhancing sustainability [20].

5. Future directions and research needs

5.1 Indirect electricity production *via* microalgal biofuels

Indirect power generation from microalgal biofuels involves many conversion methods, particularly biodiesel, biogas (biomethane), and biohydrogen, each utilising the distinctive biochemical properties and high growth rates of microalgae. These methods enhance renewable electricity production while promoting environmental sustainability and resource efficiency.

5.2 Biodiesel production and applications

Microalgae are well known for their high lipid content and have been considered as a potential feedstock for biodiesel. Lipid from microalgal biomass is extracted and converted to biodiesel by the process of transesterification, and it could be used as a fuel through a diesel generator to generate electricity. Microalgal cultures are well known for their high growth rates, high productivity and their cultivation under wastewaters, no competing with human freshwater, and arable land. By combining the cultivation of microalgae with a biorefinery concept, the utilisation of spent biomass for the production of high-value co-products can be used to strengthen economic and environmental sustainability [25]. Nevertheless, obstacles including high energy-consuming feedstock cultivation, possible microbial contamination, and expensive conversion process hinder the commercialisation of biodiesel. Current research is directed at how to increase the lipid productivity, improve the extraction technologies, and optimise the biorefinery concept to gain higher overall process efficiency and lower costs [26].

5.3 Biogas and biomethane pathways

Anaerobic digestion (AD) of microalgal biomass can generate biogas, predominantly consisting of methane, which can be utilised in combined heat and power (CHP) facilities for electricity production. Recent studies have confirmed the potential of microalgae cultivation in the operation of such co-digestion plants, with the simultaneous use of digestate nutrients and waste heat for the support of algal growth, simultaneous utilisation of wastewater, and storage of CO_2 from biogas or flue gas [27]. In addition to potential improvement in the biogas yield, this integration can lead to efficient nutrient recovery and reduction in costs. Algae-bacteria consortia have been demonstrated to have improved nutrient removal and biogas yield in marginal weather conditions. Valorisation of by-products in addition to utilisation of low-cost waste streams further amplifies the economic and environmental advantages of this approach [28].

5.4 Biohydrogen and novel energy carriers

Hydrogen production from microalgae is a promising and fast-developing bioenergy technology that is receiving increasing attention. Biological production of hydrogen can be achieved by many pathways, such as direct and indirect biophotolysis, photofermentation, and dark fermentation. Dark fermentation is also an attractive process because it can be carried out in the absence of sunlight, and continuous hydrogen can be harvested. [Fe-Fe]-hydrogenases are one of the most active hydrogen-producing enzymes found in a variety of microalgae. Nevertheless, it is still facing several obstacles, including low energy conversion efficiency and oxygen sensitivity of the enzymes that hinder its large-scale application. Recent breakthroughs include genetic modification for increasing enzyme stability, construction of microalgae–bacteria consortia, and application of nanomaterials to enhance hydrolysis. These developments are bringing the production of biohydrogen from microalgae closer to reality as an environmentally benign and economically feasible strategy [29].

5.5 Environmental and economic synergies

Microalgal biofuel systems have important environmental and social impacts, with potential for CO_2 sequestration, wastewater treatment, and the harvesting of valuable co-products, for example, proteins, pigments, and fertilisers [26]. These systems integrated with renewable electricity supply (e.g. wind or solar) could achieve an additional greenhouse gas reduction of 90% when compared with the purchase of conventional fuels. Coproduction of high-value products is a critical element to the economic viability of microalgal biofuels since the added-value product can ultimately subsidise higher costs of biomass production and processing. For instance, recent techno-economic analysis indicates that, with valorisation of co-products, the cost of algal biofuels can be brought to below $4 per GGE, thus achieving cost-performance parity with fossil-derived fuels in the near term [20].

In summary, indirect power generation by microalgal biofuels constitutes a promising and complex strategy for sustainable energy production, with current research aimed at addressing persistent technological and economic obstacles to facilitate wider commercial implementation.

6. Conclusions

Microalgal-derived electricity generation involves a wide range of emerging technologies that can be classified into direct bioelectrochemical systems (e.g. M-MFCs and photosynthetic power cells) and indirect conversion pathways based on the production of bioenergy carriers (e.g. biofuels, i.e. biodiesel, biogas, and biohydrogen). Recent progress has been 'machine-driven' by AI, energy informatics, and life cycle analysis, which enables engineers to grow, reap, and convert more effectively and more sustainably. It has become technically well developed through various approaches such as hybrid culture systems, advanced bio-flocculation and efficient lipid extraction methods.

Microalgal resources have some advantages over traditional renewables. They produce purely CO_2 emission-free electricity and, at the same time, lock up CO_2 from the atmosphere and thus can become an active tool for climate change mitigation. Compared with solar energy and wind energy, microalgae can grow on non-arable land with saline or wastewater, which will not compete with crops for food and save land and freshwater. Their high growth rates and metabolic flexibility represent a basis for the sustainable, large volume production of energy, and valuable side products towards a circular bioeconomy.

With the transition underway to decarbonised energy systems on a global scale, microalgae-based technologies are set for integration in multi-vector energy systems. They can also match with intermittent renewables to offer bioderived energy storage, support grid stability, and be located near wastewater treatment or industrial CO_2 point sources for synergistic environmental benefits. Applications are now emerging with an interest in micro-photosynthetic power cells for low energy devices, such as IoT sensors, as well as their operation in biorefinery concepts, based on simultaneous production of fuels, chemicals, and feedstocks.

Nevertheless, these technologies are not without their challenges. The large-scale cultivation, harvesting, and downstream processing costs are high, and these have been a bottleneck to commercial applications. The largely unexplored diversity of microalgal species offers untapped potential for identifying strains with improved productivity, stress tolerance, or resource-use efficiency. Areas for future research include intelligent bioprospecting, metabolic engineering for increased yields, and cost-effective, scalable bioreactor and harvesting methods. In addition, a holistic techno-economic analysis and environmental impact assessment are required to inform policy and investment decisions, at the same time balancing the economic and environmental sustainability of large-scale microalgae-based systems.

In conclusion, microalgae represent an emerging technology for future renewable energy production and are ideal for next-generation integrated solutions to both energy production, CO_2 sequestration, and resource recovery. It will take sustained innovation, cross-disciplinary cooperation, and focused investment to unlock their global potential as we transition to more sustainable energy systems.

Acknowledgements

The author gratefully acknowledges the Department of Biological Sciences at King Abdulaziz University for providing invaluable support and research facilities throughout this work. Special thanks are extended to Prof Sherif Edris for his expert guidance and mentorship, and to Dr Safiah Alhazmi and Dr Futoon Sayegh for their insightful feedback and continuous encouragement. Their collective contributions were instrumental to the successful completion of this research.

Conflict of interest

The authors declare no conflict of interest.

Author details

Wid Alrashidi
Department of Biological Sciences, Faculty of Sciences, King Abdulaziz University, Jeddah, Saudi Arabia

*Address all correspondence to: walrashidi0009@stu.kau.edu.sa

References

[1] Kim HS, Park W-K, Lee B, Seon G, Suh WI, Moon M, et al. Optimization of heterotrophic cultivation of chlorella sp. HS2 using screening, statistical assessment, and validation. Scientific Reports. 2019;**9**(1):19383

[2] Benedetti M, Vecchi V, Barera S, Dall'Osto L. Biomass from microalgae: The potential of domestication towards sustainable biofactories. Microbial Cell Factories. 2018;**17**(1):173

[3] Khan MI, Shin JH, Kim JD. The promising future of microalgae: Current status, challenges, and optimization of a sustainable and renewable industry for biofuels, feed, and other products. Microbial Cell Factories. 2018;**17**(1):36

[4] Chew KW, Yap JY, Show PL, Suan NH, Juan JC, Ling TC, et al. Microalgae biorefinery: High value products perspectives. Bioresource Technology. 2017;**229**:53-62

[5] Inuwa AB, Zeb I, Mahmood Q, Irshad U, Irshad M, Hafeez F, et al. Novel microalgae strains from selected lower Himalayan aquatic habitats as potential sources of green products. PLoS One. 2022;**17**(5):e0267788

[6] Lam MK, Lee KT. Bioethanol production from microalgae. In: Handbook of Marine Microalgae. Biotechnology Advances, Science Direct; 2015. pp. 197-208

[7] Ng F-L, Phang S-M, Periasamy V, Yunus K, Fisher AC. Enhancement of power output by using alginate immobilized algae in biophotovoltaic devices. Scientific Reports. 2017;7(1):16237

[8] Arora K, Kumar P, Bose D, Li X, Kulshrestha S. Potential applications of algae in biochemical and bioenergy sector. 3 Biotech. 2021;**11**(6):296

[9] Hussain J, Rittmann BE. Algae as a source of renewable energy: Opportunities, challenges, and recent developments. Sustainable Energy & Fuels. 2023;7(11):2515-2544

[10] Sulieman AME, Rahman EHA, Mustafa WA. Capabilities of algae to be utilized as a renewable energy source. Academic Journal of Life Sciences. 2020;**6**(3):18-26

[11] Rony ZI, Mofijur M, Hasan MM, Ahmed SF, Badruddin IA, Khan TMY. Conversion of algal biomass into renewable fuel: A mini review of chemical and biochemical processes. Frontiers in Energy Research. 2023;**11**:1124302

[12] Hu J, Meng W, Su Y, Qian C, Fu W. Emerging technologies for advancing microalgal photosynthesis and metabolism toward sustainable production. Frontiers in Marine Science. 2023;**10**:1260709

[13] Lee T-M, Lin J-Y, Tsai T-H, Yang R-Y, Ng I-S. Clustered regularly interspaced short palindromic repeats (CRISPR) technology and genetic engineering strategies for microalgae towards carbon neutrality: A critical review. Bioresource Technology. 2023;**368**:128350

[14] Subramanian S, Sayre RT. The right stuff; realizing the potential for enhanced biomass production in microalgae. Frontiers in Energy Research. 2022;**10**:979747

[15] Zhang J, Zhu Y, Hawkins TR, Klein BC, Coleman AM, Singh U, et al. Saline microalgae cultivation for the

coproduction of biofuel and protein in the United States: An integrated assessment of costs, carbon, water, and land impacts. Sustainable Energy & Fuels. 2025;**9**(7):1859-1870

[16] Włodarczyk A, Selão TT, Norling B, Nixon PJ. Newly discovered *Synechococcus* sp. PCC 11901 is a robust cyanobacterial strain for high biomass production. Communications Biology. 2020;**3**(1):215

[17] Ashour M, Mansour AT, Alkhamis YA, Elshobary M. Usage of chlorella and diverse microalgae for CO_2 capture - Towards a bioenergy revolution. Frontiers in Bioengineering and Biotechnology. 2024;**12**:1387519

[18] Ali SS, Al-Tohamy R, Al-Zahrani M, Schagerl M, Kornaros M, Sun J. Advancements and challenges in microalgal protein production: A sustainable alternative to conventional protein sources. Microbial Cell Factories. 2025;**24**(1):61

[19] Nazloo EK, Moheimani NR, Ennaceri H. Biodiesel production from wet microalgae: Progress and challenges. Algal Research. 2022;**68**:102902

[20] Office BT. Updated Algae Report Analyzes National-Scale Prospects for Converting Microalgae to Biofuels and Bioproducts Energy.Gov. US Department of Energy; 2024. Available from: https://www.energy.gov/eere/bioenergy/articles/updated-algae-report-analyzes-national-scale-prospects-converting

[21] Sundaram T, Rajendran S, Gnanasekaran L, Rachmadona N, Jiang J-J, Khoo KS, et al. Bioengineering strategies of microalgae biomass for biofuel production: Recent advancement and insight. Bioengineered. 2023;**14**(1):2252228

[22] Hasan MM, Mofijur M, Uddin MN, Kabir Z, Badruddin IA, Khan TMY. Insights into anaerobic digestion of microalgal biomass for enhanced energy recovery. Frontiers in Energy Research. 2024;**12**:1355686

[23] Ayub HMU, Nizami M, Qyyum MA, Iqbal N, Al-Muhtaseb AH, Hasan M. Sustainable hydrogen production via microalgae: Technological advancements, economic indicators, environmental aspects, challenges, and policy implications. Environmental Research. 2024;**244**:117815

[24] Ahmed SF, Rafa N, Mofijur M, Badruddin IA, Inayat A, Ali MS, et al. Biohydrogen production from biomass sources: Metabolic pathways and economic analysis. Frontiers in Energy Research. 2021;**9**:753878

[25] Casanova LM, Mendes LBB, Corrêa TS, da Silva RB, Joao RR, Macrae A, et al. Development of microalgae biodiesel: Current status and perspectives. Microorganisms. 2022;**11**(1):34

[26] Rame R, Purwanto P, Sudarno S. Sustainable energy harnessing: Microalgae as a potential biofuel source and carbon sequestration solution. Renewable Energy Focus. 2023;**47**:100498

[27] Rossi S, Carecci D, Marazzi F, Di Benedetto F, Mezzanotte V, Parati K, et al. Integrating microalgae growth in biomethane plants: Process design, modelling, and cost evaluation. Heliyon. 2024;**10**(1):e23240

[28] Omokaro G, Nafula ZS. Microalgae: Harnessing environmental and nature benefits-a comprehensive review. American Journal of Agricultural

Science, Engineering, and Technology. 2023;7(3):10.54536

[29] Li S, Li F, Zhu X, Liao Q, Chang JS, Ho SH. Biohydrogen production from microalgae for environmental sustainability. Chemosphere. 2022;**291**(Pt 1):132717